Raising Poultry the Modern Way

Updated & Revised Edition

LEONARD S. MERCIA

A Garden Way Publishing Classic

STOREY COMMUNICATIONS, INC.
POWNAL, VERMONT 05261

SF
487
.M44
1990

Cover designed by Nancy Lamb
Cover illustration by Brigita Fuhrmann
Text designed by Cindy McFarland
Illustrations on pages 9-11, 67, 89, 163, 166-169, 198 by Alison Kolesar

Printed in the United States by Capital City Press
Fourth Printing, February 1991

Library of Congress Cataloging-in-Publication Data

Mercia, Leonard S.
    Raising poultry the modern way / by Leonard S. Mercia.
        p. cm.
    "A Garden Way Publishing book."
    Includes bibliographical references.
    ISBN 0-88266-577-4
    1. Poultry. I. Title.
SF487.M44 1990
636.5--dc20   89-45738
                                          CIP

Contents

Introduction

THE INCENTIVE to write this book came in the early 1970s. It resulted from a sudden resurgence of requests for information relating to the management of small flocks of poultry. During the nearly thirty years the author was associated with poultry Extension education, never did there exist as much interest in small flocks of poultry as during the 1980s.

For years we have witnessed the development of a highly specialized poultry industry made up primarily of large, automated production and marketing units. This revolution has taken place at the expense of the small poultry flock because of diminishing profits and the problems of marketing for the small flock owners. Suddenly there developed a renewed interest in raising a few chickens, turkeys, waterfowl, or other types of poultry. Why this sudden turnabout?

The desire to produce fresh eggs and poultry for the family probably ranks high among the reasons for starting a small poultry flock. However, a few birds may also provide family members an excellent chore responsibility, hobby, or even a source of limited income. Whatever the reason, success in such a venture requires a knowledge of poultry husbandry and many other related subjects.

The need exists for a handy reference manual with complete and practical information for the small poultryman or prospective poultryman. This book is written with that end in mind. It treats practically all phases of poultry production and processing, and includes sections on chickens, turkeys, and waterfowl.

One chapter attempts to help answer the question, "Should I Raise a Poultry Flock?" and includes a discussion of the possible types of poultry projects and some of the advantages and disadvantages of each. The section on the laying flock will probably interest the larger number of readers since this seems to be the project that is of interest to most people. Much of the chapter centers around questions frequently asked of the author. It therefore covers topics such as types and sources of stock, management and feeding practices, housing, and equipment needs. It contains plans for small poultry houses, poultry furniture, a small incubator, and others. One section is devoted to meat bird production. Other sections treat turkey and waterfowl production. Information is presented on such skills as culling, debeaking, caponizing, recognition of diseases and their treatment, egg care, incubation, and egg and poultry processing. To enhance the value of the book as a handy reference, it includes lists of additional reading materials, sources of supplies and equipment, as well as the addresses of poultry specialists at the state universities and the state Poultry Diagnostic Laboratories.

Since the book does include information on more than one species, and many of the subjects discussed required references to other subjects, some repetition was unavoidable. Then, too, it was felt that repetition might be helpful to those readers who have an interest in only one section of the book or who have time to read only one section at a time. Cross references are used to prevent as much duplication as possible.

Although intended primarily as a guide and reference for prospective poultrymen and small flock owners, *Raising Poultry the Modern Way* does also have value as a text or supplement for practical poultry production courses or science projects in the schools.

Leonard S. Mercia

CHAPTER 1

Should I Raise Poultry?

THE CARE OF a small poultry flock can be a rewarding experience as a hobby, to help fill the family refrigerator or freezer, or to provide eggs and poultry for sale. Modern poultrykeeping can also be an enriching experience. With good husbandry and flock management, the poultry project can be a source of efficiently produced food. Commercially produced poultry feeds or well-formulated, home-mixed diets can be supplemented to some extent with scraps from the table, some green garden crops, and fine lawn clippings to lower feed costs. The selection of the right stock for your particular type of project will obtain optimum results whether it be laying birds, broilers, roasters, capons, turkeys, waterfowl, or others. Moreover, surplus eggs and poultry can be sold at a healthy profit.

We are currently experiencing a resurgence of interest in the production of food at home. Often the responsibility for the care of the poultry project is given to the younger members of the family. This can be a well-rounding experience for young people, fostering a basic knowledge of such things as egg care, incubation, feeding and management of growing birds and, if properly under-taken, the development of such important skills as recordkeeping and money management.

A valuable by-product of the poultry project that is frequently overlooked is the manure. Poultry manure is high in fertilizer value and is an excellent fertilizer for the garden or a valuable addition to the compost pile.

The modern commercial poultry industry is perhaps best characterized as being a highly specialized, highly commercialized agricultural industry. It is no longer a small business on many farms, but a large business on fewer and fewer farms. The keys to success in the commercial poultry business are frequently bigness and efficiency. The necessary capital investment is high; there are some pitfalls in the business, and a profitable market for large volume production is nearly essential if the enterprise is to be successful.

Small poultry enterprises with a few hundred or a few thousand birds can be profitable if the products can be marketed at a substantial margin direct to stores, restaurants, hospitals, institutional users, or direct to consumers. Even the small family flock can return a profit when products are marketed direct to the consumer at a substantial markup. These outlets, too, are not always free of

1

problems. Competition for markets is keen, and profit margins sometimes erode. Moreover, there is the problem of providing a volume of sizes, quality, and service that is often difficult for the small producer who sells to the wholesale outlets.

If you plan to sell your products and make a profit, you need to study the market potential in your area carefully. Without profitable markets any poultry enterprise is almost certainly doomed to failure.

Before starting a poultry flock, check the local zoning ordinances. Animal projects are not permitted in certain residential areas. A possible problem might involve your neighbors. Poultry flocks can bring about such problems as odors, fleas, rats, or mice. There is the possibility of complaints about noise, especially if there are adult males in the chicken flock. Nothing manages to anger neighbors more than the early morning crowing of a rooster. Good management will prevent many of these nuisances, but it is best to research thoroughly all possible problems before embarking upon your own poultry project.

Should you decide to raise poultry, it is, in most instances, a seven-day-a-week job. There are the daily management chores to be done—feeding, watering, collecting eggs, and checking the birds. Last but not least, there is the needed investment in poultry housing and equipment, as well as additional costs such as feed, birds, and supplies.

TYPES OF POULTRY PROJECTS

THE LAYING FLOCK

The laying flock is the most popular type of poultry project. Good stock properly managed will lay eggs throughout the year. Laying birds should start laying at twenty-two to twenty-four weeks of age. The length of their laying cycle is normally twelve to fourteen months. Any stock that is bred for high egg production is suitable for the laying flock. A good laying bird, under good management conditions, should lay about twenty dozen eggs during a laying cycle. Then, too, there is the possibility of an occasional meal of fowl when poor producers are culled from the flock.

MEAT CHICKENS

Specialty meat-type birds such as capons and roasters are particularly suited to small-flock enterprises. Broilers, as a general rule, cannot be grown on the small farm as economically as they can be grown commercially. However, some individuals will still choose to grow their own broilers so as to have control of the type of feed and feed ingredients the birds consume, and to have available their own freshly slaughtered birds to eat. Broilers are young chicks used for frying, broiling, or roasting, and are usually seven to eight weeks of age when processed. Commercial broilers are scientifically bred to produce meat efficiently. In the early days of the development of the broiler industry, broilers were the by-product male birds from straight-run flocks being grown primarily for replacement pullets for laying flocks. At roughly thirteen weeks of age the males were removed from the flock and slaughtered as broilers. Very few broilers are now produced as a by-product of egg production.

Roasters are three- to five-month-old chickens weighing 5 pounds or more. Capons are castrated males grown for six to seven months to weights of 7 pounds or more. The costs of raising these chickens to heavier weights are considerably higher than the costs of raising broilers. These tender-meated, well-finished birds are, however, sought after and demand a much higher price per pound in the market.

TURKEYS

Turkey production makes an excellent project for fun or profit. Although it was once thought to be hazardous and difficult to raise turkeys, this is no longer true. Thanks to the available modern-day drugs and management know-how, the turkey enterprise is no longer a particularly hazardous one. Still, certain precautions have to be observed. Blackhead disease is a possible problem, and turkeys should be reared separately from chickens to avoid problems with this disease. It is possible to raise turkeys either in confinement or on range.

A turkey project can be an excellent one for anyone, young or old. It can provide an excellent responsibility and learning experience for young people, and with good stock, good management, and good marketing, a profit can be realized. Turkeys can be raised in virtually any type of climate successfully if they receive the proper management and nutrition and are protected against diseases, predatory animals, and exposure to extreme weather conditions.

Turkey production for roasters has the added advantage of being a relatively short-term project, since it requires only about six months to prepare for the poults, grow them, process and market them, and clean the facilities in preparation for the next flock. Turkey broilers, which require only fifteen to sixteen weeks to finish, require an even shorter time period.

Fresh native turkey, properly dressed, is hard to beat for juiciness and flavor, so it is a natural for the production of food for home consumption.

WATERFOWL

Like other phases of the poultry industry, there are farms that specialize in commercial production of ducks and geese. There are also many small flocks raised to occupy the farm pond or as a hobby or sideline to other farm enterprises. Ducks and geese make an excellent hobby, and many are raised for exhibition purposes or multiplied for sale as ducklings or goslings.

Well-finished ducks and geese, are a nutritious and tasty food. Some breeds of ducks, surprisingly enough, surpass some chickens in egg-laying ability.

Rearing waterfowl is relatively easy. The young are quite hardy, and by following reasonable management and feeding programs, such a project can be a successful and profitable one. Ducks and especially geese are great foragers and can pick up a great deal of their nutritive requirements during the warm-weather months in the form of green feed, insects, and worms. In confinement they can cause damp litter conditions unless special waterers and management procedures are used.

OTHER POULTRY PROJECTS

The *American Standard of Perfection,* published by the American Poultry Association, recognizes over three hundred breeds and varieties of domesticated chickens, including bantam fowl, turkeys, and waterfowl. The *Standard* describes birds on the basis of class, breed, and variety.

The term "class" is usually used to designate groups of standard breeds developed in various regions; thus, the class names—American, English, Mediterranean, etc. The main breed differences are those of body shape and size. Feather color and color pattern, as well as comb type, distinguish the different varieties.

There are other characteristics common to most breeds within the classes. For example, American breeds commonly lay brown-shelled eggs, and those breeds within the Mediterranean class lay white-shelled eggs.

There is quite a large number of people with an interest in standardbred poultry; the interest of these "poultry fanciers" is in breeding and multiplying birds for exhibition at fairs and poultry shows. In addition to competing for prize money and trophies, they usually sell breeding stock and eggs. It provides some income and is also an interesting hobby. This project also provides eggs and poultry for the table.

Some bird fanciers also breed and exhibit pigeons and enter them in the shows. They have the opportunity to sell breeding stock, too.

Other pigeon producers raise strictly for meat production. Young squabs are sold in specialized markets for a good price—they are a real delicacy, and may provide a substantial income if grown on a commercial scale.

A number of pigeon breeders are interested in breeding racing pigeons. They usually participate in long-distance races with the hope of winning prize money.

Guinea fowl is a special breed grown for the production of meat. It provides all dark meat and brings a good price in special markets. Guinea fowl is also exhibited at poultry shows and, like other exhibition fowl, may provide the fancier with a profit from the sale of breeding stock and hatching eggs.

Game birds are also grown for a number of reasons. They provide some growers with an excellent income when produced and marketed in an area where there is a good demand. Included in this group are pheasants and quail. Breeders can sell meat birds, or breeding stock and eggs. There is a demand in some areas for birds to be used in restocking hunting farms and preserves.

The management and housing principles for exhibition fowl are much the same as those for the production of small flocks; however, breeding and selection of birds for exhibition requires a knowledge of breed and variety characteristics, genetics, and of fitting and training the birds to be shown. The housing, feeding, and management requirements for gamebirds and pigeons are substantially different from those of chickens, turkeys, and waterfowl, and space does not permit a detailed discussion of these projects in this book.

CHAPTER 2

Poultry Housing & Equipment

THE TYPE OF poultry house you'll need will vary with the type of project and how you plan to get started. For example, baby chicks and turkey poults require tighter, more comfortable housing than do older birds. If you plan to begin the project with started birds, the house may not have to be as tight or as well insulated. In warm climates the house may have open sides enclosed with poultry netting. It may be equipped with adjustable awnings for use during inclement weather. Waterfowl require very little in the way of housing as adults; however, young ducklings and goslings, though relatively hardy, will need good housing during the early brooding period.

The type of house will also depend, to a certain extent, upon the management system you use. Some differences in design and size might be needed to accommodate different types of equipment and furniture. If cages are utilized for laying flocks, the building construction and design must be considered to insure that the equipment fits in the building. Material-handling chores such as feeding, cleaning, watering, and gathering eggs should also be considered when planning the poultry house.

The poultry house need not be elaborate or expensive. An old building can frequently be remodelled to accommodate a small flock of poultry. Sometimes a pen built within a large existing structure will function quite well for small flocks. Remodelling of old buildings for large production units is seldom advisable; although some old structures are suitable for the production of broilers, pullets, or turkeys, most are not readily adapted to the efficient handling of birds. Material-handling chores for young growing birds are not so critical, but it is an important consideration for large laying projects.

There are basically two types of poultry housing. One type is designed for floor management systems, the other for cage systems.

Floor housing, also called litter or loose housing, allows the birds free access to the poultry house or pens within the house. The advantage of the floor system is that it permits flexibility in the use of the house. It can be adapted for brooding and growing or managing most any type of poultry flock including broilers, roasters, capons, layers, turkeys, or waterfowl. This system is well

7

suited for the small farm or family flock, and most of our consideration will be given to this type of housing.

A word about cages. This is the system most frequently used by commercial egg producers today. There are several reasons for its popularity. Floor space requirement per bird is less, thus reducing per bird cost of housing. Cage systems are readily adapted to labor saving in materials handling. The use of automatic feeders and waterers, the belt collection of eggs, and manure-handling devices are a natural with cages. Another important advantage of cage units is that eggs do not have to be collected as frequently because of the egg roll-a-way feature. Cage housing and management systems permit the care of more birds per person than does the floor system. For the most part we find that cage facilities offer better control of parasites and eliminate many of the troublesome litter-borne diseases that we sometimes find with floor birds. Most breeding farms still use the floor systems of housing for the mating of breeders and the production of fertile eggs. Breeding cages are available, and can be used successfully for light breeders.

A poultry house, or poultry pen, must provide a clean, dry, comfortable environment for the birds throughout the year. If young birds are to be brooded in the house, it must be built well enough to permit a comfortable environment and efficient use of fuel. It must be tight enough to prevent drafts on the birds and help maintain a uniform temperature economically. Poultry facilities should have floors that can be easily cleaned and disinfected. Most poultry pens or buildings should provide for ventilation to control moisture in the pens, remove gases, and conserve or dissipate heat. In the northern climates the houses should have insulation in both the sidewalls and ceilings. Piped-in water is desirable but may create problems in severely cold climates, especially in the small flock situations where body heat is not sufficient to keep the pen temperatures above freezing. It is possible to correct this particular problem through the use of electric heat tapes and water warmers. Houses should have artificial light to provide the right lighting program for both layers and growing stock. The poultry house should be strong enough to withstand high winds and snow loads if necessary. The house plan shown (Figure 1) is for a small flock and provides room for storage of feed and supplies.

SELECTING A SITE FOR THE POULTRY HOUSE

When selecting a site for a poultry house consider the soil drainage, air movement, location of the dwelling, and water supply. Remember, there is the possibility of odors, flies, rats, and mice. Good soil drainage assures dry floors,

Figure 1. **POULTRY HOUSE PLANS**

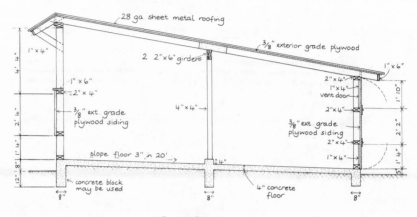

Cross Section

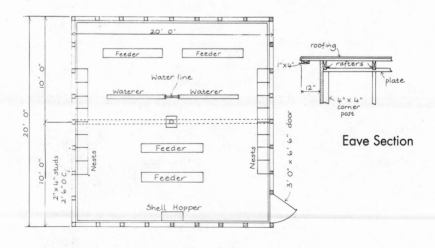

Eave Section

Plan

Figure 1 (continued)

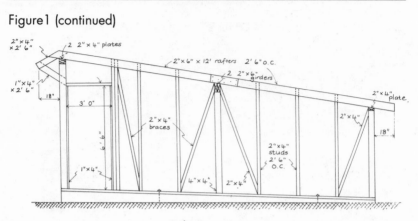

Side Framing

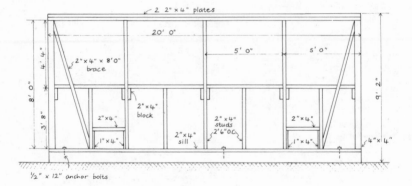

Front Framing

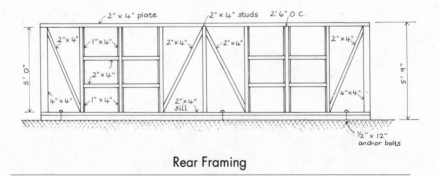

Rear Framing

Figure 1 (Continued)

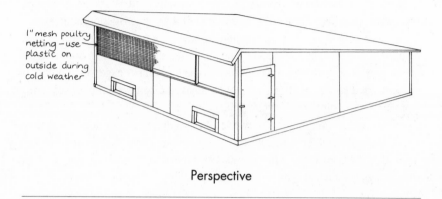

1" mesh poultry netting – use plastic on outside during cold weather

Perspective

which will help prevent wet litter, dirty eggs, disease, or other problems.

Locate the poultry house where the prevailing summer winds will not carry odors to the dwelling. The site should be large enough to provide for expansion, if this should seem advisable in the future. The location of access roads, electric lines, future buildings, and yards should also be considered.

A site on relatively high ground with a south or southeast slope and good natural drainage is desirable. Siting a poultry house at the foot of a slope where soil or air drainage is poor, or where seepage occurs, is unwise.

If the house is located on a hillside, the site should be graded so as to carry surface water away from the building. In some instances tile drains will be necessary to adequately carry the water away from the foundation of the building.

In areas of severe winter weather, most of the windows should be in the front of the building and the house should face south to take advantage of the sunlight. The pen will be warmer, the litter drier, and the birds more comfortable. The closed or back side of the house should provide maximum protection against the northwest wind. Where the prevailing winter storms are from the west, however, orient the house facing east. Wider houses are often placed with their long dimensions north and south to provide lighting on both sides. If the house is oriented with the front window area to the south, solar heat can be controlled by the width of roof overhang to provide shade in the summer months when the sun is high on the horizon, and permit sunlight in the pens in the winter when the sun is low.

SPACE REQUIREMENTS

The size of the poultry house will depend upon the type and number of birds to be housed as well as the management system to be used. Various age groups and species of birds require different amounts of floor space for optimum results. If the house is designed for a laying house, and a cage management system is to be used, the size of the building will depend upon the dimensions of the cages, the type of cage, and the number of birds to be housed per cage. In addition to space needed for the birds, there should always be some additional space for storage of feed, supplies, and equipment.

Floor space requirements vary according to the type of project, the size or age of the birds, and the management system in use (Table 1). It is important to have adequate floor space to prevent such problems as cannibalism, poor growth or poor egg production, and morbidity or mortality.

INSULATION

Removal of moisture from the building is one of the main problems of poultry housing. The average moisture content of freshly voided poultry manure is 70 to 75 percent. The moisture content of the manure varies with the type of feed, feed ingredients, temperature in the building, and even the type of bird. Waterfowl, for example, tend to produce wetter droppings than do chickens and turkeys. Then, too, respired moisture and the moisture of the air brought into the building by the ventilation system must also be removed from the building.

Insulation will provide optimum bird comfort and avoid excessive moisture problems if the heat is adequate. In most areas of the United States the poultry house should be insulated. The purpose of insulation is to conserve heat in the cold climates and to keep out the heat in warm climates. It should be pointed out that light-colored reflective roof and wall surfaces are helpful in keeping the pens cooler in warm climates. Since warm air carries more moisture than cold air, insulation serves to conserve heat in cool climates not only to provide birds warmth, but also to permit removal of moisture from the building. It is important to keep the litter dry and to prevent buildup of ammonia in the house. Excessive ammonia can be detrimental to the birds and uncomfortable for the operator.

It was once thought that only housing in the northern climates need be insulated. The development of the so-called controlled environment house for high density operations has changed this philosophy, and we now find more of this type of housing being built in the south as well as in the northern climates, particularly for cage management systems. Controlled environment houses are

TABLE 1
FLOOR SPACE REQUIREMENTS FOR POULTRY

Type of Bird	Age of Bird	*Floor Space (sq. ft.)
Chicks	0-10 weeks	.8–1.0
Chicks	10-maturity	1.5–2.0
Layers	Brown egg	2.0–2.5
Layers	White egg	1.5–2.0
Layers	Meat-type Breeders	2.5–3.0
Broilers	0-8 weeks	.8–1.0
Roasters	0-8 weeks	.8–1.0
Roasters	8-12 weeks	1.0–2.0
Roasters	12-20 weeks	2.0–3.0
Turkeys	0-8 weeks	1.0–1.5
Turkeys	8-12 weeks	1.5–2.0
Turkeys	12-16 weeks	2.0–2.5
Turkeys	16-20 weeks	2.5–3.0
Turkeys	20-26 weeks	3.0–4.0
Turkeys	Breeders (heavy)	6.0–8.0
Turkeys	Breeders (light)	5.0–6.0
Ducks	0-7 weeks	.5–1.0
Ducks	7 weeks maturity	2.5
Ducks	Breeders (confinement)	6.0
Ducks	Breeders (yarded)	3.0
Geese	0-1 week	.5–1.0
Geese	1-2 weeks	1.0–1.5
Geese	2-4 weeks	1.5–2.0
Geese	Breeders (yarded)	5.0

*Many factors determine the floor space requirements including the type of management system, the type of house, the number and kind of bird, the climate, and even the management the birds receive.

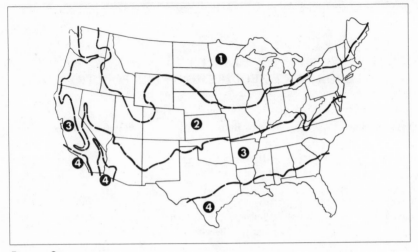

Figure 2.

windowless structures with artificial light and fan ventilation.

Climate influences poultry house design, especially for the small flock. The farm building zone map (Figure 2) shows the four basic climatic zones in the United States, based on January temperatures and relative humidity.

In the northern part of the United States the small flock (less than a few hundred birds) is not without its special problems. Even the well-insulated

TABLE 2
RECOMMENDED MINIMUM INSULATION
VALUES FOR POULTRY HOUSES BY ZONES

LOCATION	R-VALUE	
	Walls	Ceilings
Zone 1:		
Colder parts	8-10	10-15
Warmer parts	6	12
Zone 2	5	10
Zones 3 and 4	2	5

Insulation value (R-value) is defined as the number of degrees difference in temperature between the inside and outside surfaces of a wall that will permit 1 British thermal unit (B.T.U.) of heat to pass through 1 square foot of the wall per hour.
    Source: USDA

poultry house may not provide optimum conditions during severe winter weather unless artificially heated. It is essentially impossible or highly impracticable to insulate buildings well enough to conserve the heat given off by a small number of birds. Frozen waterers, frozen pipes, or even frozen combs may result, causing a drop in egg production. Cold buildings also cause excessive feed consumption. Well-insulated pens within large existing buildings may be one answer to some of the above-mentioned problems, especially if other animals are housed in the building and help provide some of the warmth within the structure. Houses that are well protected from the wind are also much easier to keep warm and comfortable in winter weather. Table 2 gives the recommended insulation values for poultry houses in the different climatic zones. The resistance of insulation values of various building materials is shown in Table 3.

In zone 4, laying houses can be built with wire walls, which can be covered with curtains during cool or windy weather. In zone 3, uninsulated houses with large openings in the front are frequently used. These are usually equipped with windows or plastic curtains for use in cold or stormy weather.

VENTILATION

The purposes of ventilation are to provide bird comfort by removing moisture and ammonia from the building, to provide an exchange of air, and to control the environmental temperature in the pen.

There are two types of ventilation systems commonly used in poultry houses: the natural or gravity system, and the forced-air system.

The gravity system may utilize windows, special slot inlets, or flues to provide air movement. The incoming cool air replaces the warmed, moisture-laden air going up through flues in the ceiling or through the top of opened windows. If flues are used in cold climates, they must be insulated to prevent condensation of moisture which drips back into the pen. Flues must also extend well above the roof to provide the necessary draft to remove the warm, moisture-laden air. This approach to ventilation works satisfactorily in an insulated house where inlets or windows are properly adjusted. The gravity method of ventilation has been replaced gradually as the size of poultry operations have increased. The need to open and close windows with changes in temperatures and wind conditions is a very time-consuming chore. Maintenance of the windows is also time consuming and costly.

Forced-air ventilation gives the best control of air movement. Fans are used to move out the warm, moisture-laden air and ammonia, and bring in cool air for bird comfort through adjustable slots located in the walls near the ceiling. Fan capacity is determined by the type, age, and number of birds in the house. For

TABLE 3
R-VALUES OR INSULATION VALUES OF
VARIOUS BUILDING MATERIALS

MATERIAL	THICKNESS IN INCHES	RESISTANCE OR R-VALUE
Air space	3/4—4	.91
Blanket insulation (glass or rock wool)	1	3.70
Blanket insulation (glass or rock wool)	2 1/2	9.25
Blanket insulation (glass or rock wool)	3	11.10
Building paper		Negligible
Cinder block	8	1.73
Concrete	10	.80
Concrete block	8	1.11
Fill insulation		
shavings	3 5/8	8.85
shavings	5 5/8	13.70
sawdust	3 5/8	8.85
sawdust	5 5/8	13.70
fluffy glass fiber, mineral or rock	3 5/8	13.40
Homosote	1/2	1.22
Insulation board (typical fiber)	1/2	1.52
Insulation board (typical fiber)	25/32	2.37
Plywood	3/8	.47
Sheathing and flooring (softwood)	3/4	.92
Shingles (asphalt)		.15
Shingles (wood)		.78
Siding (drop)	3/4	.94
Siding (lap)		.78
Surface (inside)		.61
Surface (outside)		.17
Window (single glass)		.10
Window (double glass in single frame)		1.44

example, a house to be used for two thousand growing or laying birds would be designed with a fan capacity of from 500 to 10,000 cubic feet per minute (cfm). This flexibility of air movement can be accomplished with either multiple fans, two-speed fans, or a combination of both. It permits a small amount of air movement for young chicks or for cold weather conditions, and a larger volume of air movement for older birds or warm weather conditions.

In cold climates 5 cfm would be too much air movement for bird comfort, so the ventilation rate is cut back to a lesser rate by the use of thermostats which control the fans. In cold climates the optimum temperature for all but young birds under six to seven weeks of age is approximately 55° F. Fans should not be allowed to cycle on and off at frequent intervals. Fan cycling occurs when pen temperatures are cooled down too quickly by overventilating. The maximum amount of moisture is not removed from the pens, and the birds are exposed to uncomfortable temperature conditions when this occurs. Fan cycling can be prevented with proper thermostat settings, and by restricting the air inlets in cold weather, to maintain a rather definite relationship between pen temperature, incoming air, and exhausted air. In the warm months the inlets are opened wide to permit a maximum flow of air for removal of moisture and ammonia and for cooling the birds. Houses for small flocks can usually be ventilated satisfactorily through the windows.

MANAGEMENT SYSTEMS

There are essentially three types of management systems used for the care of poultry. The one most widely used, until a few years ago, was the floor or litter management system. This is still the most popular method for small flocks, breeder flocks, growing birds, broilers, roasters, turkeys, and waterfowl. This system permits the birds freedom of the entire pen or building. Litter materials such as wood shavings, ground corn cobs, sawdust, sugar cane, or finely chopped straw are used on the floor to insulate the floor for bird comfort and to absorb moisture. Litter also helps to control disease and enhances egg cleanliness in laying flocks. In addition to the necessary feeding and watering equipment, the system may utilize nests or roosts or dropping pits, depending upon the type of bird housed. A variation of this confinement management system utilizes a fenced yard adjoining the house or pen. Yards are useful for small flocks, particularly where floor space is limited. A grass sod in the yard can also provide some forage for the birds. The use of the yard depends upon the age and type of bird housed, climate, and other factors.

Another type of management system utilizes slats or wire mesh as the floor. This system has been used for laying birds, turkeys, and waterfowl. The entire floor is slats or wire. No litter is used and roosts are not required. The bird's droppings fall through the wire or slats into a pit and are generally cleaned out periodically or when the birds are removed. A variation of this method incorporates either slats or wire and a litter section. The feeders and waterers are usually placed over the wire or slat sections, and the fecal material drops into a shallow or deep pit. The pit and the slatted or wire section are usually located in the center of the pen with a litter section on each side. This system has several advantages, in that the birds spend approximately 75 percent of their time on the slats or wire

TABLE 4
EQUIPMENT REQUIREMENTS (PER 100 BIRDS)

Bird Type	Age (weeks)	Brooders Hover-type	Brooders Infrared Lamp	Feeder	Waterer
Chicks— Broilers, Roasters, Capons, & Pullets	0–2	700 sq. inches	2	1 feeder lid to 7 days 100" of trough or 3 hanging	20" trough or 2 one-gal.
	2–6	700 sq. inches	2	200" of trough or 3 hanging	40" trough or 3 one-gal.
	6–maturity	—	—	300" of trough or 4 hanging	96" of trough or 1 automatic
Layers	22–market	—	—	300" of trough or 4 hanging	96" trough or 1 automatic
Turkeys	0–2	1200 sq. inches	2–3	1 feeder lid to 7 days 200" trough or 3 hanging	20" trough or 2 one-gal.
	2–6	1200 sq. inches	2–3	400" trough or 5 hanging	96" trough or 1 automatic
	6–market	—	—	480" trough or 6 hanging	120" trough or 1 automatic
Ducks & Geese	0–2	1200 sq. inches	2–3	1 feeder lid to 7 days 100"of trough or 3 hanging	20" trough or 2 one-gal.
	2–4	1200 sq. inches	2–3	200" of trough or 5 hanging	50" trough or 3 automatic
	4–market	—	—	200" of trough or 4 hanging	50" trough or 3 automatic

Note: A trough feeder or waterer 2 feet long and open on both sides provides 48 inches of space.

sections eating and drinking. Therefore, the majority of the droppings go into the pits, helping to keep the litter sections dry. This greatly simplifies the job of ventilating the building and results in cleaner litter and cleaner nests and eggs. The same types of feeders, waterers, and nests are used for litter-floor houses as for the slat- or wire-floor houses.

The third type of management system utilizes cages. As mentioned earlier, cages have become the most popular means of managing market egg birds; most of the large commercial laying houses are now equipped with wire cages. Cages have sloped floors so eggs roll to the front for easy collection. Since the birds can't set on the eggs, they need not be collected as frequently. Cages are also coming into use to brood pullets to be housed in laying cages. Thus far, attempts to raise meat birds in cages haven't been too successful.

FURNISHING THE POULTRY HOUSE

Furnishings for the poultry house depend upon the type and size of the project. If it is a growing project the major requirements will be brooding equipment, feeders, and waterers. A laying bird project will require feeders, waterers, nests, roosts, and a broody coop for birds that go broody. If laying cages are used, no nests, roosts, or broody coops are needed.

The equipment need not be fancy. Feeders and waterers should be designed to service the birds efficiently with a minimum of waste or spillage. There must be enough equipment to give each bird an equal chance to utilize it. The amount of equipment required varies with the type of bird, age, and size (Table 4).

Equipment can be purchased from local feed and farm supply outlets or from mail order houses and farm equipment concerns (see Sources of Supplies and Equipment section). For the small poultry operation much of the equipment can be homemade or purchased secondhand. Feeders and waterers are included with the cage equipment.

FEEDERS

Feeders should be large enough to supply the flock's needs for a day or more without wasting feed. Proper feeder design is an important consideration in avoiding feed waste. The use of an anti-roost device, such as a reel or a spring-loaded wire on top of the feeders, will help to prevent waste. A lip on the side of the feed hopper will prevent birds from beaking out feed. Feed hoppers should never be filled more than one-third to one-half full to avoid feed waste. The size,

FEEDERS

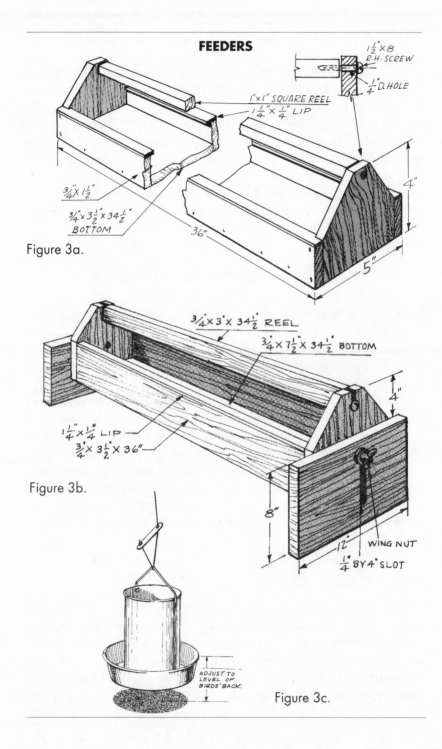

$1\frac{1}{2}"\times 8$ R.H. SCREW

$\frac{1}{4}"$ D. HOLE

$1"\times 1"$ SQUARE REEL

$1\frac{1}{4}"\times \frac{1}{4}"$ LIP

$\frac{3}{4}"\times 1\frac{1}{2}"$

$\frac{3}{4}"\times 3\frac{1}{2}"\times 34\frac{1}{2}"$ BOTTOM

36"

4"

5"

Figure 3a.

$\frac{3}{4}"\times 3"\times 34\frac{1}{2}"$ REEL

$\frac{3}{4}"\times 7\frac{1}{2}"\times 34\frac{1}{2}"$ BOTTOM

4"

$1\frac{1}{4}"\times \frac{1}{4}"$ LIP

$\frac{3}{4}"\times 3\frac{1}{2}"\times 36"$

8"

12"

WING NUT

$\frac{1}{4}"$ BY 4" SLOT

Figure 3b.

ADJUST TO LEVEL OF BIRDS' BACK.

Figure 3c.

FEEDERS (CONTINUED)

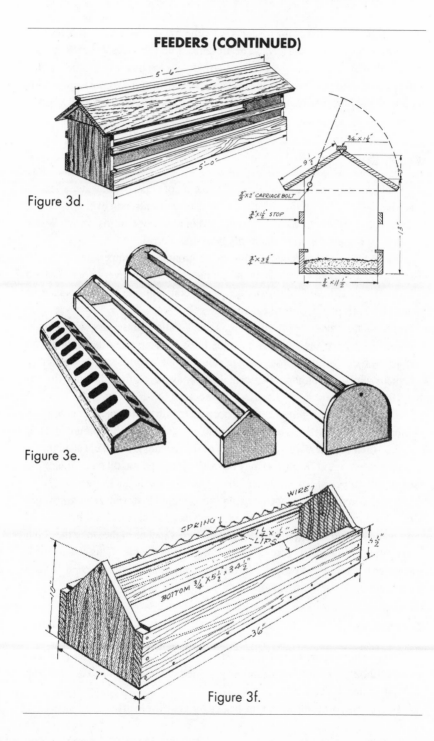

Figure 3d.

Figure 3e.

Figure 3f.

height, and construction of the feeder is very important if feed waste is to be kept to a minimum. To minimize feed waste, troughs should be made so that the height can be adjusted as the birds grow. The alternative is to change to a larger feeder as the birds become larger. Less feed is wasted if the lip of the feeder is level with the top of the bird's back. Some of the feeders commonly used and wooden feeders you can build are shown in Figures 3 through 3f.

WATERERS

Birds of all ages should have access to plenty of clean, fresh water. Bear in mind that water is one of the cheapest sources of nutrients, and over one-half of the bird's body is water. Eggs are composed of about two-thirds water. Water is a vital ingredient for all of the bird's body functions.

Water consumption depends upon the environmental temperature, age, and species. Laying birds will drink about 2 pounds of water for every pound of feed consumed. In extremely hot weather water consumption may amount to 4 pounds for each pound of feed eaten.

For young chicks the waterers can be 1-gallon glass jars with plastic bottoms. These are easy to clean. As the birds become older these should be exchanged for larger metal fountains or troughs. Older birds can be watered in open pans, pails, or troughs. The containers may be equipped with a float to make them automatic.

If running water is piped to the house you may want to consider buying an automatic water fountain. The cost is usually not excessive, the amount of labor required is reduced, and a constant supply of fresh water is available at all times. A homemade waterer (Figure 4) may be made from a gallon oil can. Figures 4a and 4b show types of waterers commonly used for small flocks of poultry.

A fountain should be placed on a platform (Figure 5) or water stand, covered with a 1 x 2 mesh welded wire. The platform can be made of 2 x 4 material 30 to 36 inches square. For best results, raise it 3 to 4 inches higher than the depth of the litter. Better still, place it over a drain if available. This arrangement will lessen the amount of wet litter surrounding the fountain and also help to keep litter material out of the fountain.

BROODERS

If day-old birds are to be started, some type of brooder (heat source) will be needed. Enough brooder space must be provided so that every bird can get under the heat source as needed.

Brooders may be the hover type heated by gas, oil, electricity, wood, or coal.

WATERERS

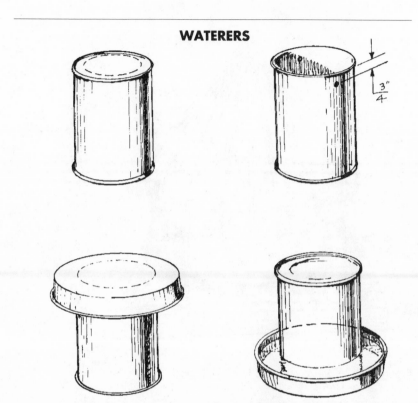

Figure 4. Homemade waterer made from a gallon oil can and a pan.

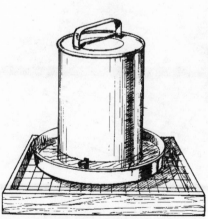

Figure 4b. Chick waterer.

Figure 4a. Waterer for older birds.

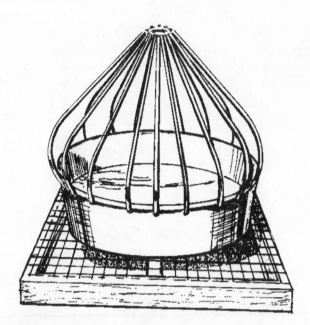

Figure 4c. Waterer suitable for laying birds.

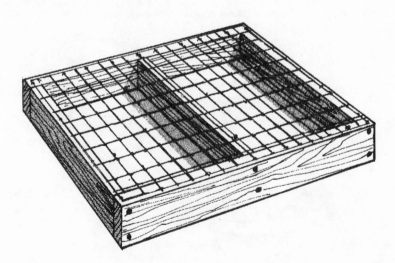

Figure 5. Wire water stand.

Infrared lamps work well for small flocks and have the added advantage of enabling you to observe the chicks at all times. Electric brooders are more expensive to operate than most other types.

Figure 6 shows an infrared brooder with guard. Figure 6a is a plan for a homemade electric brooder for a small flock. The typical hover-type (Figure 6b) reminds one of a flying saucer. It is usually suspended from the ceiling by chain or cable.

ROOSTS

Roosts are not essential but are recommended for most small laying flocks. They should not be used for broilers or other meat birds, as they may cause blisters. Roosts are used for turkeys but not waterfowl. Pullets should have 6 to 8 inches of roost space per bird. If the layers are expected to roost they must be trained at an early age, or it will take considerable time to train them to use the roosts in the laying house.

There are several types of roosts. One of the more common types is the dropping pit with perches on the top (Figure 7). This type of roost is designed to accumulate droppings underneath for several months. This, naturally, keeps a large amount of the moisture out of the litter. If properly constructed and screened in around the top and sides, birds will be prevented from getting into it. The nesting material and litter stay cleaner. Dropping pits are usually located either in the center of the building or at the rear of the building—these are normally the more comfortable areas in the house. The dropping pit should be designed to facilitate culling and so that it can be moved easily for cleaning. In larger units the dropping pits should be narrow enough to facilitate easy catching or culling of birds at night.

Another roosting method uses a dropping board. The typical dropping board is usually mounted on the wall at the height of about 2½ feet. The open space under the dropping board provides additional floor space for the birds. The roosts are located above the dropping board and the perches are hinged to the wall so they can be raised when the dropping board is cleaned. This method is not as desirable as the dropping-pit type and is not recommended unless floor space is at a premium.

The roosts or perches should be of 2 x 2 stock, rounded or bevelled on the upper edges to prevent injury to the breast and feet. For small breeds, allow 8 inches of perch space per bird. Large breeds should have 8 to 10 inches of perch space. Roost perches are usually spaced 12 to 15 inches apart.

Figure 6. Infrared brooder.

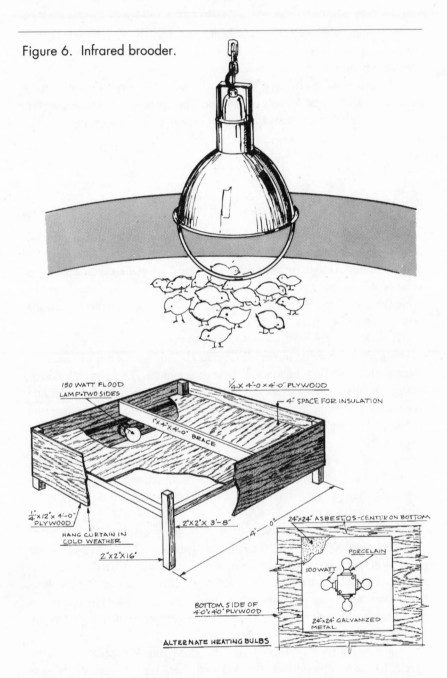

Figure 6a. Homemade brooder.

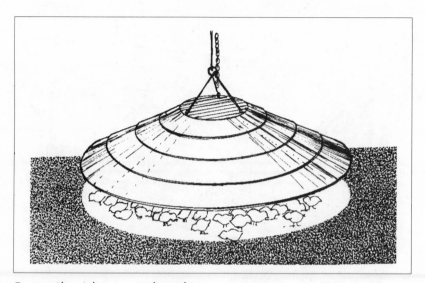

Figure 6b. A hover-type brooder.

NESTS

An adequate number of well-designed nests with clean nesting material should be provided for the laying hens. They can be either individual or community-type nests. One individual nest, or 1 square foot of community nest, should be provided for each four laying birds. Individual nests (Figure 8) should

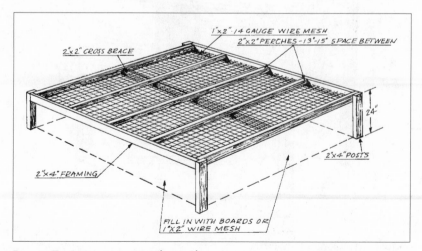

Figure 7. Dropping pit with perches.

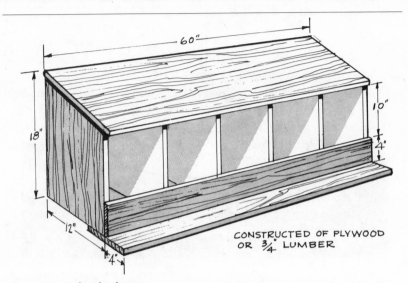

Figure 8. Individual nests.

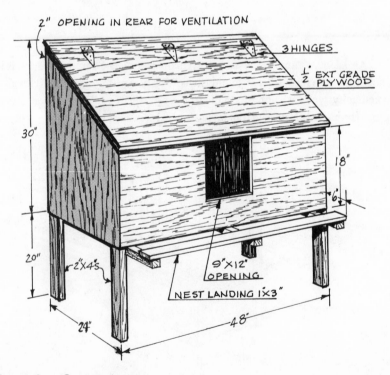

Figure 8a. Community nests.

be at least 1 foot square and 1 foot high. Community nests (Figure 8a) can be almost any size but must be provided with at least two openings, 9 x 12 for every 20 square feet of nest space. As an aid to getting the birds into the nests, darken by covering two-thirds of each nest entrance with a cloth flap. Place a landing board below the nest openings to provide easy access to the nests. The nests should be located approximately 2 feet from the floor or surface of the litter.

Commercially made nests are also available from agricultural supply houses and specialized equipment dealers. Some of these people offer a community or individual type with a roll-a-way feature. After the eggs are laid, they roll out to the front or back of the nest into an egg tray where they can be easily gathered. A word of caution about this type of nest: It's excellent in theory, but birds must be trained to use it and, more often than not, a large percentage of the eggs are

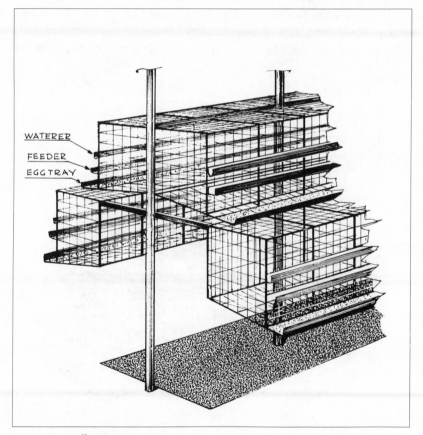

Figure 9. Full stair-step cage.

laid on the floor rather than in the nest.

Nests should be cleaned frequently, as the nest material becomes dirty. Since several hens use the community-type nest at one time, ventilation of the nest is very important. For this reason provide a 2-inch-wide opening at the top of the nest to allow heat to escape.

CAGES

Laying birds may also be housed in wire cages. Cages can be located in an open building or in closed fan-ventilated housing. As mentioned earlier, cages simplify the management of layers, but birds in cages are more susceptible to the effects of extreme weather conditions. Therefore, they must be protected from wind, cold, and particularly, from hot weather.

Cages (Figures 9–9b) are usually constructed of 1 x 2 welded wire; the floor is constructed with a slope of approximately 2 inches per foot, so that eggs will roll out onto the egg tray. Feed and water is provided in troughs usually located

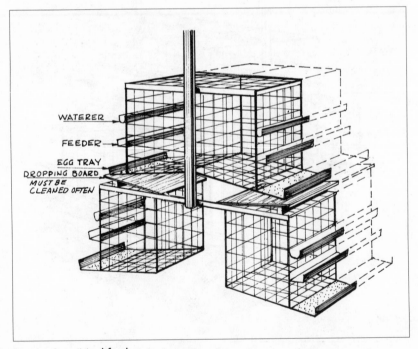

WATERER

FEEDER

EGG TRAY

DROPPING BOARD
MUST BE
CLEANED OFTEN

Figure 9a. Modified stair-step cage.

on the outside of the cage. Water cups located inside the cage are also sometimes used.

There are several types of cages available, most of which are quite satisfactory. They come in several sizes but should not exceed 20 inches in depth for best results. Cages 12 inches wide, 18 inches deep, and 16 inches high are very popular. Research has indicated that three brown-egg birds or four Leghorn-type birds can be placed in a 12 x 18 cage with good results.

Several types of cages are satisfactory. They include the stair-step, the double-deck with dropping boards, the colony, and the single-deck. Stair-step cages are available in two arrangements, either full or modified. Some commercial producers are also moving toward a triple-deck or even a four-deck modified stairstep-cage system to enable them to house more birds in a given area.

The use of cages cannot be justified in many of the very small laying flock enterprises; however, there are situations in which they can be used to good advantage. We have helped plan several small cage units for flocks of three or four hundred birds, utilizing an existing building or pen space more efficiently than could have been accomplished using a floor system. In these small units the birds are usually watered automatically, but feeding, egg collection, and cleaning are manual chores.

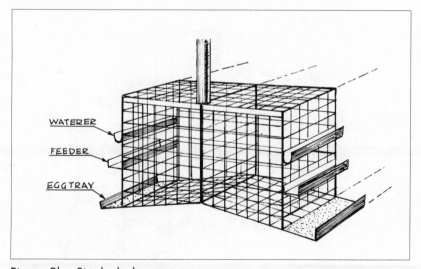

Figure 9b. Single-deck cage.

CHAPTER 3

Starting the Laying Flock

THE LAYING FLOCK can be started in a variety of ways. You can set eggs under a broody hen or in a small incubator. Day-old chicks or started pullets can be purchased. One questionable method of starting is to purchase second-year layers. These are birds that have laid for a year. Production is not as good for second-year layers; egg quality may be poor; and they may stop laying and go into a *molt* (shed old feathers and grow new ones) soon after moving.

One of the simplest and best ways of starting the laying flock is to buy started pullets. This avoids the need for equipment and the care involved in incubating eggs and brooding chicks. You can buy started pullets from commercial growers at nineteen to twenty weeks of age. Sometimes it is possible to buy a few surplus pullets from a local poultryman in your area. Pullets should be debeaked to avoid feather picking and cannibalism. Ten to fifteen birds will normally provide the average family with enough eggs. Unless surplus eggs can be marketed easily you should not produce more than the family needs. Bear in mind that a healthy, well-managed laying flock can produce twenty-one to twenty-two dozen eggs per bird in a laying year.

CHOOSING THE BEST CHICKS

Before buying birds, decide whether you want white or brown eggs. Egg shell color does not affect food value, but it does influence the market price, since there is a definite preference in various areas of the country. In most sections of the United States, white-shelled eggs sell for slightly more than the brown-shelled eggs. In the New England area, or so-called Boston market area, brown eggs are preferred and the situation is reversed.

Birds that lay white eggs usually weigh in the vicinity of 4 to 4½ pounds at maturity. Representatives of these breeds and varieties are the Leghorns, first generation Leghorn strain crosses, or hybrids—all of which are good egg producers but not so good for meat.

The medium-weight breeds, American Breeds, lay brown eggs. The mature weight of these birds is usually between 5 and 6 pounds, thus making them more desirable as meat birds. Most of the medium-weight, brown-egg producing layers are crosses of American breeds, such as the Barred Plymouth Rock and the Rhode Island Red. There are several of these crosses, all of which are good egg producers. Several pure breeds such as the Rhode Island Reds, White Plymouth Rocks, and New Hampshires can be used as dual purpose breeds, that is for meat and eggs. However, they may not do as well as the crosses bred especially for meat or egg production.

Stock bred specifically for high egg production or for meat production is best suited for the home flock. To get the project off to a good start, purchase well-bred, healthy chicks or started pullets. One mistake frequently made is the purchase of chicks from distant hatcheries. Sometimes the conditions to which chicks are exposed when being transported over long distances are less than desirable. They may be chilled or overheated, leading to early losses and a poor-producing flock. It is usually best to purchase chicks or started pullets from a reputable nearby hatchery or producer that has stock bred for efficient production.

TRAITS TO CONSIDER

Traits to consider when purchasing chicks or pullets include livability, early feathering and feed efficiency, freedom from disease, rate of growth, egg production, and egg quality. Much valuable information can often be obtained by talking with poultry producers in your area. They can usually tell you from experience which breeds and strains have the above characteristics and the most reputable hatcheries and pullet growers from which stock can be obtained. Another means of evaluating stock is to consult the random sample egg production and meat production test reports, which are conducted in several states. These tests are designed to test the entries of production stocks for the various economic production traits. The test reports provide information to help evaluate the performance of laying stock or meat stock offered for sale by the participating breeders and hatchery personnel.

One of the first steps toward getting disease-free chicks is to buy the stock from hatcheries that blood test their breeders for pullorum-typhoid. Hatcheries operating under the National Poultry Improvement Plan do blood test their breeding flocks for these diseases. Both of these diseases are passed from the infected hen to the chick through the egg.

If a nearby hatchery has stock with good production capabilities and is disease free, it is best to get your chicks locally to minimize the shipping time and

Rhode Island Reds

Single Comb White Leghorns

Sex-Sal Layer
(Popular brown-egg New England crossbred)

Source: *Poultry Tribune*, Mount Morris, Illinois

avoid the possibility of prolonged exposure to extremes of temperature and poor handling. It is usually much easier to get adjustments from a nearby hatcheryman in the event that problems arise. Chicks should be ordered at least four weeks in advance of the date you would like to start them, and started pullets should probably be ordered close to six months before they are needed. Your county Extension Agent or Extension Poultry Specialist can help you select sources of stock (see the Directory of Cooperative Extension Offices in the back of this book). For a list of hatcheries participating in the National Poultry Improvement Plan, write to the United States Department of Agriculture, Beltsville, Maryland 20705.

BUYING YOUR CHICKS

The importance of selecting the right stock should be re-emphasized here. After deciding on the breed or strain you want, be sure to buy healthy stock from hatcheries with a U.S. pullorum-typhoid clean status. Buy the best chicks you can buy. Cheap chicks may cost you more in the end than those that cost most in the beginning. Poor results with the growth or livability can soon erase any savings you may have made in the purchase of chicks.

Be sure to buy chicks bred for the purpose for which they are to be used. If you are interested in starting with a flock of laying birds, get chicks from a source known for having birds with high egg production records. There are great differences among types of birds, insofar as egg size, egg production, feed efficiency, livability, and many other economic traits are concerned.

In the New England area, production birds that lay brown eggs are primarily the sex-link crosses. Several crosses of this type are offered for sale, known by various names. They are called sex-linked crosses because the breeds that are mated result in male chicks of one color and female chicks of another color. Day-old chicks can be sexed by feather color, thus eliminating the cost of vent sexing.

When buying chicks for laying stock the usual practice is to purchase sexed pullets. This is because the cockerels of the best laying stocks are apt to be poor meat producers; it is usually not wise to try to grow them for meat.

Only birds bred for meat production should be raised as broilers, capons, or roasters. Stock that is bred for the production of meat has the ability to grow rapidly with a minimum of feed consumption. Feed consumption represents at least 60 percent of the total costs of producing poultry meat, thus feed cost becomes an important consideration. Also, birds bred for meat tend to mature with a better finish, are more tender, and juicier. Then, too, the meat bird stocks are primarily white feathered, which results in a much better-dressed appearance in the absence of dark pinfeathers.

WHEN TO START CHICKS

The time to start chicks may depend upon the type of housing you have and the type of brooding equipment available. Chicks started in the winter commence laying in the early summer, but may go into a neck molt and take a vacation the following winter. On the other hand, pullets, which are hatched in late winter or early spring, start to lay in the summer when egg prices are normally highest. Those raised late in the spring will not start laying until late fall when egg prices are usually lower.

Probably the best time for most small-flock producers to buy their chicks is in late March, April, and May, especially for those in the northern climates. Chicks started at that time do not need cold-weather brooding facilities. Moreover, the pullets will start laying in early fall and will continue laying for a full twelve-month cycle.

CHAPTER 4

Brooding & Rearing
Young Stock

BROODING FACILITIES can affect the degree of success one has in growing chicks. As indicated earlier, an expensive building is not necessary, but the facility should be built to brood chicks efficiently.

One of the features of a good brooder house is structural soundness to withstand high winds and snowloads in cold-winter areas. It should be of tight construction, with sufficient insulation against drafts, and easy to maintain a uniform temperature economically. It should have a ventilation system that controls moisture and gases, yet still is able to conserve or dissipate heat. Lastly, it is desirable to have some degree of light control in the house.

EQUIPMENT

The main equipment needs include the feeders, waterers, and brooders. Roosts may be used for replacement pullets.

FEEDERS

Commercial or homemade feed hoppers are satisfactory for brooding chicks. Probably the commercial hoppers are easier to clean and cause less feed waste. Round hanging feeders are excellent for small flocks. They are easily adjusted to accommodate birds of different ages and to avoid feed waste. Since they have storage space they don't have to be refilled as frequently as the hopper-type feeders, and the feed stays clean.

WATERERS

It is possible to make water fountains using oil cans or juice cans in combination with tin plates. Make two holes on opposite sides of the can about ¾ inch from the lip. Fill the can with water, place the plate on top, and flip the can and plate over to create a waterer that maintains its own water level. See Figure 4 for details on making this waterer.

A small wire platform underneath each fountain helps to keep the water clean, keep the shavings out of it, and also prevents the litter from becoming too damp around it.

Probably the best watering device for chicks is the 1-gallon fountain. Commercial poultry producers frequently use the gallon fountains in combination with automatic waterers. They gradually move the gallon fountains close to the automatics. They are thus able to convert to the automatic waterers as early as ten days of age. Once the chicks are drinking well from the automatic waterers, the fountains can be removed gradually and additional automatic fountains added until they have three to four 8-foot waterers per one thousand birds.

In small flock situations, two 1-gallon waterers are usually good for one hundred birds for up to two weeks. From the second through the sixth week, there should be at least two 2- or 3-gallon waterers, and from eight to ten weeks, one 5-gallon waterer per one hundred birds.

BROODERS

To provide the necessary heat some type of brooding device is required. Infrared lamps are an excellent source of heat for brooding small flocks of chicks (Figure 6). The initial cost of equipment is relatively small, and the lamps are excellent for supplying heat to the chicks, particularly in the late spring, summer, or early fall months. The infrared lamp is suspended approximately 18 inches above the litter. One 250-watt bulb is sufficient to brood up to seventy-five chicks. Even though the number of chicks may be small, it is wise to use at least two bulbs in case one fails.

Many small flocks are brooded with homemade brooders utilizing electric light bulbs as the heat source. It is possible to construct simple hovers (canopies) of plywood and canvas, or some similar material, which can be suspended from the ceiling or supported by legs (Figure 6a).

Other brooders manufactured commercially are fired by oil, gas, electricity, wood, or coal. These have a hover to keep the heat down at floor level (Figure 6b). Such brooders are excellent for flocks of one hundred chicks or more. Some of them are rated at a capacity in excess of 750 chicks. The initial cost of these units is considerably more than either the infrared lamp or the homemade brooder. Each chick needs a minimum of 7 square inches of hover space or its equivalent.

Each type of brooder has certain advantages. The infrared bulb has the advantage of enabling the operator to see the chicks at all times. Where the hover-type brooder is used it is more difficult to observe the chicks. On the other hand, the hover-type brooder is best for cold-weather brooding and is less expensive to operate.

The type of equipment selected will depend, to a certain extent, upon the cost and availability of fuel and the extent to which it is to be used. Whatever brooder stove you use, it should have the capacity to heat adequately and a means for controlling the temperature accurately.

ROOSTS

As noted in Chapter 2, roosts should not be used for broilers, roasters, or capons. In fact, few people use roosts even for growing replacements. Those who want their birds to roost in the laying house feel that it is necessary to train the pullets to roost in the brooder house. If roost perches are used, they should be enclosed with poultry netting to prevent the chicks from coming in contact with the droppings. They should be constructed to provide easy access for the birds—often they are slanted from the floor to the wall much like a ladder.

PREPARATION OF THE BROODING QUARTERS

Getting the chicks off to a good start is very important. Probably no other part of the enterprise deserves more planning and advance preparation than does the starting of baby chicks. The modern brooder house should be clean and disinfected or fumigated at least two weeks before the chicks arrive. If this is not done, the chicks may be exposed to certain poultry diseases which can lead to mortality and poor results. First sweep or wash down cobwebs, dirt, and dust, then use a good disinfectant. There are a number of good commercial preparations available. They should be used according to the manufacturers' recommendations. Some of them require that the house stand idle for a period of time and be thoroughly aired before the chicks arrive. Failure to observe these precautions may cause the chicks to have severely burned feet and eyes.

After the brooder house has been cleaned and disinfected, it should be allowed to dry thoroughly before putting in new litter. When the floor is dry, cover with 2 to 4 inches of litter material. The litter material serves to absorb moisture and to insulate the floor for comfort of the birds. Several different materials may be used for litter. Wood shavings are among the most commonly used, whenever available. Other materials include sawdust, or shavings and sawdust, rice hulls, sugar cane, peanut hulls, and ground corn cobs.

When chicks are extremely hungry upon arrival they may try to eat the litter before learning to eat feed. If you have reason to believe that the chicks have been hatched for two or three days when they arrive, it may be a good idea to put paper over the litter for the first three or four days until they learn to eat. When chicks are received soon after hatching, it is not necessary to do this.

The brooders should be started a day or two before the chicks are due to arrive. This will assure that the equipment is operating properly and is adjusted to the correct temperature. Use a chick guard to confine the chicks to the source of heat. The chick guard keeps the chicks confined to a given brooder and also prevents migration and overcrowding of some brooders if more than one brooder is used in the brooding facility. It will also help to prevent drafts on the chicks under the brooder. A corrugated cardboard guard, approximately 12 inches high, is good for this purpose during cool weather. For warm-weather brooding the guard may be made of poultry netting. It should form a circle around, and about 3 feet from, the brooder. The brooder guard can be extended outward after a few days. It is usually removed at the tenth day or so, depending upon the weather and conditions in the brooder house.

Fill the feeders and waterers several hours before the chicks arrive. The water will be at room temperature when the chicks arrive, and encourage them to drink. The feed trays and water fountains should be spaced uniformly around

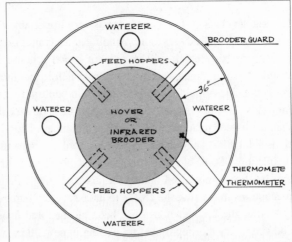

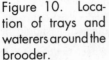

Figure 10. Location of trays and waterers around the brooder.

Figure 11. **BROODER MANAGEMENT**

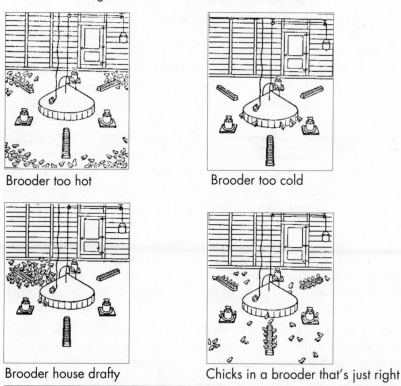

Brooder too hot

Brooder too cold

Brooder house drafty

Chicks in a brooder that's just right

the brooder and close to the hover (Figure 10). One method frequently used to get chicks to eat, when first put under the hover, is to place a small amount of feed on either new egg filler flats or on inverted chick box lids (the type of lid without the slots or holes for ventilation). Chicks instinctively peck at anything at the same level as the surface upon which they are standing. Therefore, they will learn to eat more readily if feed is provided in this manner for a day or two. Chicks that are slow to catch on to the fact that feed is in front of them are attracted by the noise made by the others that have found the feed and are pecking on the cardboard. This gets the chicks off to a good start and eating well.

BROODER MANAGEMENT

Brooding temperature is very important. The recommended temperature at the start is 90° to 95° F. A rule of thumb to remember: Reduce the temperature 5° each week until the chicks no longer need heat. It is often stated that the best thermometer, to gauge the most comfortable temperature, is to watch the chicks themselves (Figure 11). After the chicks reach seven days of age this is probably true. If the chicks huddle close to the heat source you can be reasonably sure that the operating temperature is too low. If, on the other hand, the birds are located in a circle, way outside the heat source, you should assume that the temperature is too high. During the day the chicks should be evenly distributed around the entire brooding area, with some of the chicks underneath the heat source. Temperature readings should be taken at the outside edge of the brooder at chick level.

One important factor for the successful brooding of chicks is the maintenance of good litter conditions. When litter conditions get out of hand and become too wet, disease problems can result. Wet, dirty litter can harbor many disease organisms that affect poultry. In the case of replacement pullets, it is desirable to have a certain amount of moisture in the litter (30 to 35 percent) to enable sporulation of oocysts and the subsequent development of an immunity to coccidiosis during the growing period. However, an excessively wet litter, especially during warm periods, can bring about a clinical case of coccidiosis and result in what we term a coccidiosis break, which requires treatment.

One of the common problems experienced with meat birds is breast blisters—swellings or external sores on the skin of the breast—which seriously detract from the dressed appearance of the carcass. These are quite common in flocks of capons and roasters. Excessively wet, caked, or dirty litter is frequently blamed for breast blisters. Since perches are not used for broilers or meat birds, the birds must bed down in the litter. It is important that the litter depth be maintained and that it is kept dry and fluffy enough to cushion the body and avoid all contact with the floor. Furthermore, it is essential that it be reasonably clean to avoid soilage of the breast feathers, skin irritations, and breast blisters.

GENERAL FEEDING PROGRAMS

The recommended modern feeding programs for meat birds and laying bird replacements vary widely. Broilers and other meat birds are fed a high-energy diet throughout the growing period. In the production of meat birds, the goals are rapid growth, heavy weights, and efficient feed conversion. Replacement pullets, on the other hand, are started on medium- to high-energy starter rations for the first six to eight weeks; then they are changed to a grower diet to accommodate their changing dietary needs.

Meat bird chicks and replacement pullets reared on litter are normally started on a diet containing a coccidiostat. A higher level of coccidiostat is used for meat birds to prevent coccidiosis, a disease common to chickens reared on the floor. The coccidiostat is usually fed throughout the growing period to meat birds to protect them against a coccidiosis outbreak and to obtain maximum growth and efficiency. Replacement pullets are fed a lower level of coccidiostat in the starter diet. This enables them to develop an immunity to coccidiosis by six to eight weeks of age, and this immunity protects them throughout the remainder of the growing period and the laying period.

Diets for meat birds, particularly broilers and turkeys, are usually somewhat better fortified with vitamins and growth-promoting factors. They sometimes contain antibiotics. The protein level of meat bird diets, at least during the first several weeks, is higher than for replacement birds. Another difference is that feeding programs for meat birds do not include scratch grains during the growing period. Occasionally, heavy meat birds are fed corn the last two weeks of the growing period to improve their finish just prior to dressing or marketing. Frequently a special finishing diet is used for this purpose. Some growers of meat birds prefer the crumbles or pellets to the mash form because of less feed waste and increased feed consumption. However, crumbles and pellets cost more, and there is also a tendency for birds fed on crumbles, and particularly pellets, to feather-pick or become cannibalistic.

Some broiler or roaster feeds contain feed additives that must be removed several days prior to slaughter. The withdrawal period varies with the type of drug—for the last several days the birds are fed on a special feed without additives. Some drugs used for the treatment of diseases are administered through the feed. Definite withdrawal periods are required for these drugs prior to slaughter. Always follow the recommendations of the feed company printed on the feed tag.

FEEDING AND MANAGEMENT OF LAYER REPLACEMENTS

Replacement chicks for layers are started on a diet very similar to the broiler diet; however, the energy and protein content are usually slightly lower. The protein content is 20 to 21 percent. Chicks to be raised for layers are normally fed a starter for six to eight weeks and then changed to a grower diet.

As with other animals, the growth rate diminishes as the birds become older. Therefore, the protein requirement is less later in the growing period and the grower diet is formulated with somewhat less protein than the starter.

Several feeding programs are used from the six to eight week period through housing. The two basic systems are the all-mash diet with approximately 15 percent protein or a mash and scratch diet. For a number of years it was felt that it was best to full-feed growing replacements. Today, more and more growers are using some form of feed restriction to delay sexual maturity, yet permit the development of a healthy pullet. Other growers are using lighting programs to delay sexual maturity. No doubt the simplest and best procedure for the small flock manager is to full-feed the growing pullets. When feed is restricted, there must be adequate equipment and good management to avoid problems.

WHY DELAY SEXUAL MATURITY?

Birds that commence to lay too early or that are too fat at the onset of production frequently do not perform as well as slower-maturing birds. They are inclined to lay fewer and smaller eggs, and are more prone to prolapse of the uterus and higher mortality.

Birds on a full-feed program that are grown in the off-season (under conditions of increasing day-length) should receive a controlled lighting schedule to delay sexual maturity. This method, if done correctly, usually produces the most economical gains and more uniform birds with fewer social problems or vices.

There are several restricted feeding programs. One of the fairly common ones is the use of a 15 or 16 percent protein diet of medium energy. The exact diet depends upon the type of bird. Usually some physical restriction is used. It can be a skip-a-day program, or a specific amount of feed can be fed daily depending upon the weight and age of birds. The restriction requires attention to certain details. For example, it requires about 50 percent more feed space for a restricted feeding program to assure that all birds get feed. Then, too, restricted programs tend to increase feather pulling and cannibalism, and may increase the nervousness of the flock.

One feeding program still used to a considerable extent is a combination of the grower concentrate with a grain such as oats. After eight weeks the oats can be gradually added to the grower diet until proportions of approximately one-half grower concentrate and one-half oats is reached. Some producers permit the birds to free-choice feed the mash and oats. Grit should be used when a whole grain is fed. A hard, insoluble grit, not a soluble grit such as calcite crystals, should be used. Use a medium or pullet-size grit to approximately twelve weeks of age and the hen size grit, or coarse, for the remainder of the growing period. Grit must be present in the gizzard to enable birds to grind the fibrous material so it can be utilized and absorbed in the intestines.

Other restriction programs now receiving some attention for delaying sexual maturity in pullets are high-energy, low-protein-type diets and low-lysine diets. Good pullets can be grown on most of these programs. Check with your feed company and the breeder of the stock you purchase to determine which system he recommends.

RANGE REARING

Under some conditions range rearing, or yarding, can still be recommended. It may be possible to save some of the cost of feed and still grow good healthy pullets. Practically all broilers and most meat birds are reared in confinement, so as to have absolute control of the feed intake, more rapid and efficient growth, and better finish. As mentioned earlier, rapid growth and early maturity are not desired in replacement birds, so range rearing or yarding may be used for replacement pullets.

Range rearing was once a very popular method of growing young stock during warm weather. There is no question that excellent pullets can be grown on range. Good range-reared birds usually have excellent shank color, beautiful feathering, and are inclined to be big-framed birds having health and vigor. Why, then, have many producers gone to confinement rearing?

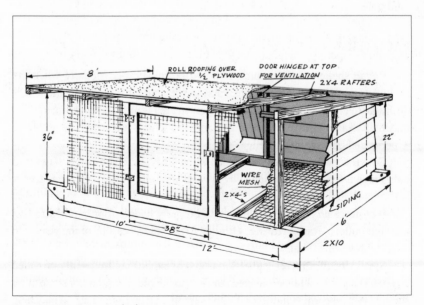

Figure 12. Range shelter.

DRAWBACKS

Losses from predators and the increased labor requirements in feeding and caring for the birds are probably two of the biggest reasons for getting away from range rearing of pullets. Predators such as foxes, coons, and even hawks and owls can be a real problem where birds are kept on range. Then there is the expense of building fences and range shelters, and their upkeep. If the birds are to be yarded you can perhaps utilize the brooder house for shelter, thus avoiding the need for special range shelters. The choice of whether to range or confine birds will depend upon a number of factors, such as the size of flock, the labor situation, and the land and housing facilities available. Figure 12 shows a range shelter suitable for about one hundred pullets.

FEED

A good range properly managed will yield a good supply of forage for growing pullets. Some of the most satisfactory grasses and legumes for poultry range include ladino clover, Kentucky bluegrass, and brome grass. This type of range permits feeding of a grower diet on a restricted basis and makes the birds forage for a portion of their nutrient intake. The range should be kept clipped to be most palatable and kept in a sanitary condition, that is, without bare spots and mud holes.

FACILITIES

Range shelters and equipment should be moved occasionally when the ground around them becomes bare, so as to keep the birds out of the mud. It is desirable to change range areas so that a given area is not used every year. This may help to avoid disease problems. Capillaria worms are one of the problems experienced on wet muddy range.

Covered feeders are recommended when birds are fed on range. They will help prevent feed loss or spoilage from wind and rain. A covered range feeder is shown in Figure 3d.

It requires an acre of good range to properly handle four hundred to five hundred pullets. Range rearing is limited to the warm seasons of the year or to areas with warm climates.

Small flocks of layers, pullets, roasters, or capons are frequently provided a yard. The yard is a fenced-in area attached to the brooder house or some other shelter. A yard permits more birds to be kept in a given floor area and may help prevent management problems such as cannibalism or feather picking. Its use in the cold climates is pretty much restricted to the warm months.

Where yards are used, the feeding and watering equipment is usually left in the house. If covered feeders are used, some of the feeding may be done in the yard.

The same hazards may exist with yards as with range or pasture. Disease may be a consideration where the same yard is used year after year. Ideally the yard area should be moved each year. Predators may also be a problem. Layers permitted to yard may hide their eggs in areas outside the house. In wet weather the eggs may be excessively dirty. It is usually best to confine the laying flock for optimum results.

VACCINATION

Several vaccines are available to prevent outbreaks of some of the troublesome poultry diseases. Some of the common immunizations are for infectious bronchitis, Newcastle disease, laryngotracheitis, fowlpox, and cholera. In some areas, all of these diseases are a problem. Newcastle and bronchitis can be problems anywhere, and fowlpox is prevalent in many areas. In northern New England we don't see fowlpox very often and do not recommend vaccination for it. The same has been true for laryngotracheitis. Vaccination programs should be established on the basis of those diseases prevalent in the particular area. Commercial producers routinely vaccinate for Newcastle and bronchitis in all areas of the United States and, in some areas, also vaccinate for fowlpox and cholera on a routine basis. Vaccination recommendations in various areas, therefore, will differ, and the individual program should be developed to handle the disease situation in the particular area.

Some of these vaccinations come early in the chick's life; the Marek's vaccination, for example, is administered at day of age, usually at the hatchery. Incidentally, it is advisable to purchase chicks that have been vaccinated for Marek's disease. Marek's disease did, for many years, take a tremendous toll on growing birds. With the advent of the Marek's vaccine, mortality has been cut to a very negligible rate. Other vaccinations are administered at various times during the growing period and may require booster shots during the laying period.

There are two basic types of vaccines: those that produce a temporary immunity and those that produce a solid or permanent immunity, which will protect the birds for the remainder of their lives. The virus vaccines producing a temporary immunity are usually of low virulence or, in some cases, have no apparent affect on the birds when administered. The virus vaccines producing a permanent immunity are of a more severe virulence, and therefore will produce more marked or serious symptoms of sickness.

There are two varieties of vaccines on the market with regard to method of application. There are those applied to individual birds and those applied to the flock on a mass basis. Two of those applied on a mass basis are vaccines for Newcastle and infectious bronchitis. These may be applied as a spray, dust, or in the drinking water. Those vaccines used on a mass basis are mild vaccines of low virulence.

Some vaccines are not permitted in certain states. Vaccination of small flocks is not recommended in many areas. Contact your Poultry Diagnostic Laboratory or Poultry Specialist at your state university. Local industry service people and county Extension Agents can also be helpful.

Because there are so many factors to consider, and so many variable conditions from farm to farm, no set vaccination program can be recommended here that will best suit all farms. The program should be planned to meet the needs of the given area. It should be planned well in advance of the brooding and rearing season, and be scheduled so that all the birds that are to be placed in laying houses will have been vaccinated well before the time they are ready to lay. This will give them sufficient time to fully recover from the stress of vaccination and avoid any carry-over effects there might be on egg production or egg quality.

FEATHER PICKING AND CANNIBALISM

Both feather picking and cannibalism are rather common vices which can develop in the brooder house and may be carried on into the laying house. A similar type of problem is toe picking, which can start in a flock of chicks soon after they are put down under the brooder.

The exact cause of these vices is not always evident. Some factors that are thought to contribute to them are poor nutrition, overcrowding, or other faulty management practices. Overheating the birds, lack of floor, feed, and water space, or even overlighting are all thought to be factors. Sometimes the problem will occur even with apparently good management.

At the first sign of feather picking, action should be taken to stop it immediately. When blood is started the birds frequently pick each other apart and losses can be severe. In some flocks, the problem can be corrected by a change in management—providing more feed, water, or floor space, better ventilation, cutting back on the light, or any number of things that will improve the birds' comfort or change environmental conditions.

In the case of small flocks, anti-pick ointments applied to the birds have been somewhat successful in curtailing the problem. Plastic and metal devices known as "specs" can also be used. Specs are mounted on the top beak and are designed to obstruct the bird's forward vision. Sometimes these devices catch on the poultry furniture and come off or partially unattach.

DEBEAKING

The most widely used, and probably the most satisfactory method of cannibalism control, is to debeak the birds. Many producers debeak their birds routinely to prevent picking problems.

Chicks can be debeaked at day old at the hatchery, using a precision debeaking machine. The beaks tend to grow out again. To be safe, most producers debeak again before the birds are put into the laying house. Unless the job is done properly, debeaking at day of age can result in chick mortality due to starve-outs.

A number of producers are now debeaking at seven to nine days of age. Special adaptors for debeaking machines make this a precision job. If done properly, this usually holds and the job need not be done again. Debeaking can

Figure 13. Properly debeaked pullet.

Figure 13a. Properly debeaked adult bird.

be done at any age to correct a picking problem. However, it is best done before the birds start to lay and should be done by sixteen weeks of age.

Debeaking involves cutting off slightly more than one-half of the upper beak and blunting the lower beak. The upper beak should be shorter than the lower beak, making it difficult for the bird to grasp feathers or skin. A temporary debeaking job can be done in the absence of a debeaking machine. This is done by removing a small portion of the upper beak with nippers or dikes. The beak soon grows out and the procedure needs to be repeated.

A debeaking machine has an electrically heated blade that cuts and cauterizes at the same time to avoid bleeding. Properly debeaked growing and adult birds are shown in Figures 13 and 13a.

LIGHTING PROGRAMS FOR YOUNG STOCK

The first three days to a week, the chicks should be provided twenty-two hours of light. During this time they get used to their surroundings and learn where the heat, feed, and water are located.

After that initial period careful attention should be given to the lighting schedule for chicks grown for layers. Birds are very responsive to light. This response to light is an excellent example of the interaction of the nervous system with the various endocrine glands. The eye receives the light stimulus, causing the release of certain hormone-releasing factors from the hypothalamus of the brain, which in turn causes hormonal secretions in other glands. The endocrine glands play an important role in many body processes including growth, rate of sexual maturity, and rate of egg production.

If growing birds are exposed to an increasing daylight length, particularly towards the end of the growing period, they will mature sexually at too early an age. The result will be smaller eggs, which are less valuable than the larger sizes. Total egg production may be lower, and there may be a high incidence of prolapse of the uterus and other problems. The rule of thumb to follow is never expose laying birds to a *decreasing* light period, and never expose growing pullets to *increasing* daylight length.

At the equator, daylight length remains constant, so lighting schedules are not much of a factor in pullet growth and production. North and south of the equator, daylight length varies seasonally: the further from the equator, the greater the difference.

Several programs have been suggested and used for growing pullets to prevent early sexual maturity and its related problems. Lighting programs are

TABLE 4a

CONVENTIONAL (OPEN-SIDED) HOUSES

Growing Period

—Start chicks on 22 hours of light for the first week of age. Reduce light and light intensity to 18 hours in week two and 16 hours in week three. Light intensity during growth should be 5 to 10 lux (0.5-1 footcandle) at bird level.

—From the 4th week through 16 weeks Harco Sex-Link commercial pullets should be grown on a constant day length according to the appropriate light requirements for month and latitude in the following table. Light intensity during this period should not exceed 10 lux (1 footcandle) at bird level.

Production Period

—At 16 weeks of age, increase light intensity to 20 lux (2 footcandles) at bird level. This level should be maintained throughout the production period.

—Where constant day length in the growing period was less than 14 hours, increase day length by 1 hour in week 17 and 1 hour in week 18. Continue to increase day length by 30 minutes per week until a maximum day length of 16 hours is reached.

—Where constant day length in the growing period was 14 hours or greater, increase light one hour in week 17 and add light in 30 minute increases until a 17 hour day length is attained.

—Keep clocks on Standard Time and divide day length increases evenly between morning and evening.

recommended based on the type of housing used. In windowless (environmentally controlled or light-tight) houses, the suggested lighting program usually seeks to maintain a continuous light period after the first few weeks, using 1 footcandle of light at bird level.

One such program, for example, calls for twenty-two hours of light for the first week, eighteen hours the second week, and sixteen hours the third week. Beginning with the fourth week and continuing through the sixteenth week, the birds are kept on an eight- to ten-hour day. At seventeen weeks through twenty weeks, the light is increased one hour weekly; and from twenty-one weeks, the light is increased thirty minutes per week until the day length reaches fifteen hours.

One brown-egg breeder recommends the following lighting program for conventional or open-sided houses.

Pullets grown in open-sided or windowed houses are exposed to natural daylight conditions and are treated somewhat differently. If the birds are hatched

TABLE 4b
CONSTANT LIGHT PROGRAM IN GROWING PERIOD
CONVENTIONAL (OPEN-SIDED) HOUSES

HATCH DATE		HOURS OF LIGHT IN GROWING PERIOD		
		LATITUDE		
NORTH OF EQUATOR	SOUTH OF EQUATOR	0°–29°	30°–39°	40°–45°
January 15	July 15	13½	14½	14½
February 15	August 15	14	15	15½
March 15	September 15	14	15	15½
April 15	October 15	13½	14	14
May 15	November 15	12½	13	13
June 15	December 15	12	11½	11½
July 15	January 15	11½	10½	10
August 15	February 15	11½	10	9
September 15	March 15	11½	10	9½
October 15	April 15	11½	11	10½
November 15	May 15	12	12	12
December 15	June 15	13	13	13½

Example: A flock hatched in Thailand (15° N latitude) on April 15. Maximum natural light during growth will be 13½ hours. Provide 13½ hours of light from 4 weeks of age through 16 weeks of age.

during the time of year when daylight length will be decreasing during the growing period (April through July in the northern hemisphere), the pullets can be reared under natural lighting conditions. If the chicks are hatched during the August through March period, special lighting programs must be used—either a step-down program or a constant lighting program.

The step-down program is frequently used in the New England area for August through March hatches. The length of day twenty-four weeks from the date of hatch is determined. Six hours are added to the length of day at that time and the birds are started on that amount of light. The total light period is then reduced fifteen minutes per week throughout the growing period and, at twenty-four weeks of age, the birds will be on a fourteen-hour day. Then add thirty minutes over the next two weeks until total day length reaches fifteen hours. It should be noted that the light periods in the programs for open-sided or windowed houses is that total of natural daylight supplemented by artificial light.

COSTS OF GROWING PULLETS

Many factors affect the costs of production. The most flexible cost will be the cost of feed. Where home-grown grains are available or a range rearing program is used, substantial savings may be made and the costs will vary from those shown in Table 5.

TABLE 5
ESTIMATED COST OF RAISING PULLET REPLACEMENTS
TO TWENTY-TWO WEEKS*

ITEM	BROWN-EGG PULLET	WHITE-EGG PULLET
Chick cost	$.60-1.00	$ .60-1.00
Feed cost @ $.10 per lb.	2.00	1.60
Brooding, litter, and misc.	.20	.20
TOTAL	$2.80-3.20	$2.40-2.80

Assumptions: Mortality 5%, 20-pound feed consumption for brown-egg birds, 16-pound consumption white-egg birds full-fed. Costs will vary depending upon quantities purchased, feed program, feed prices, and other factors.

*Interest on average investment, labor, and overhead (depreciation, insurance, and taxes) are costs of producing commercial pullets not normally considered in small-flock situations. The estimated costs of these items are: interest—13.6–15.2 cents; labor—25.2 cents; overhead—25.5–28.4 cents.

CHAPTER 5

Managing the Laying Flock

PREPARING THE LAYING HOUSE

THE LAYING HOUSE should be thoroughly cleaned between flocks—this means the removal of manure, and brushing down the cobwebs and dust from walls and ceilings. Wet cleaning is best, but dry cleaning is satisfactory. Wet cleaning involves scraping all the caked material from the floors and walls, then thoroughly scrubbing and rinsing with a garden hose or steam cleaner. To dry-clean, scrape the caked material from the floors and walls and sweep down all surfaces to remove the dust and cobwebs. Any repairs that are needed should be done at this time.

After the house is cleaned it should be sprayed with a disinfectant. The equipment should also be cleaned and disinfected at this time. Some of the common disinfectants include creosol solutions, hypochlorites, and the phenol or carbolic acid preparations. Use disinfectants according to the manufacturer's recommendations.

When the house has been disinfected, the roosts, dropping boards or pits, and the nests should be painted with carbolinium or a red-mite paint. It is advisable to treat the floors with carbolinium once each year. Carbolinium is not only a disinfectant but a wood preservative, and it makes future cleaning much easier. A similar cleaning and disinfecting program should be used for the brooder house and equipment.

A word of caution: Carbolinium is an excellent material but must be used correctly. Allow a minimum of two weeks for the house to dry and air out before the birds are put in. The material can be injurious to the birds' feet and eyes. After the house has had an opportunity to dry, the litter can be added.

LITTER

Provide approximately 6 to 8 inches of clean, dry litter material. Although the availability and cost are factors in determining litter to be used, some of the more commonly used materials are shavings, sawdust, a combination of shavings and sawdust, ground corn cobs, and sugarcane. These are all excellent litter materials.

To help maintain the litter in a satisfactory condition, it is recommended that a built-up litter be used. Built-up litter should be started early in the fall before cold weather sets in. Start with 4 to 5 inches of clean litter and gradually add to it. Fresh litter is added until a depth of 9 to 12 inches is reached. It is usually not necessary to change the litter during the laying year. Removal of the wet spots from time to time is usually all that is required.

Built-up litter insulates the floor and provides warmth for the birds. Due to the decomposition and fermentation processes, heat is actually produced in the litter. It also absorbs moisture from the feces.

The same litter should not be used for more than one year. Infestations of internal and external parasites can result. Then, too, it is necessary to remove the litter to properly clean the house.

When warm weather arrives, it is the practice of some producers to replace the built-up litter with new litter. This provides cooler, more comfortable conditions during hot weather. In cold weather it is usually necessary to stir the litter regularly to keep it from packing and to permit it to aerate and dry.

FEEDING THE LAYERS

Laying birds require a diet containing from 16 to 18 percent protein. Actually, the requirements vary during the laying cycle. Protein intake needs to be higher during the early laying period, because that is when egg production peaks and the birds are still growing. As egg production diminishes, the protein requirement decreases.

Protein is expensive, so commercial producers frequently phase-feed their layers, that is, use three different protein levels during the laying period. They may start with an 18 percent diet and, at approximately four months of production, reduce the level to 16 percent. When the layers fall below 60 to 65 percent production, the protein content of the feed is dropped to 15 percent.

Many factors affect feed intake of the birds. With changes in feed consumption, protein intake changes, so feeding layers is not a simple task if you are aiming for optimum production. Some of the factors that affect feed intake include:

1. Management factors such as feed hopper and floor space.
2. Environmental temperatures, both warm and cold.
3. Calorie content of the diet.
4. Variations in egg production.
5. Flock health or stress factors.

Most small-flock owners prefer to use a simple feeding program, one that utilizes one type of feed and can be full-fed to the layers. In these situations a 16 to 17 percent all-mash diet is normally used.

ALL-MASH FEEDING

The simplest and most foolproof approach to feeding layers is to feed a complete all-mash laying ration and keep it in front of the birds at all times. It is less bother, is adaptable to mechanical feeding, and provides a more nearly balanced diet for the layers, assuming there is enough feeder space available.

GRAIN AND MASH FEEDING

Mash is a mixture of finely ground grains. It may be formulated so as to provide all the nutritive requirements of the birds or formulated to be fed with grain or other supplements. Grit refers to hard, insoluble materials fed to birds to provide a rinding material in the gizzard. Grit may be useful when birds are fed high-fiber feed ingredients, to aid in their digestion. Small stones or granite particles are good grit materials.

Some producers still prefer to use a grain and mash system of feeding. Grains are frequently fed in the form of so-called "scratch feed," containing corn, wheat, and oats. It is often thought to be somewhat easier to maintain a high feed consumption with the grain and mash systems. Grains are also well liked by the birds and they eagerly consume them when they are provided.

Some producers feed a large part of the daily allowance of grain before the birds go to roost. During cold weather this gives them a reserve supply of energy for warmth and will help keep them more comfortable while on the roost during the night. Grains are high in energy and are digested more slowly than mash, thus providing more warmth over a longer period of time. Grit must be provided when hard grains are fed.

The portions of mash to grain vary with the protein content of the mash and also the type of layer. Usually a 20 to 21 percent mash is kept before the birds at all times. Light breeds such as Leghorns, on a grain and mash program, are fed equal amounts of grain and mash, or perhaps slightly less grain than mash. Heavy breeds have a tendency to become too fat if they receive too much energy, and therefore may receive only 40 percent grain and 60 percent mash. In some instances, if a high-energy type mash is used, it is necessary to decrease the amount of grain to 30 percent of the total diet. It is important to prevent laying birds from getting too fat. Excessive fat is not conducive to high egg production.

One of the problems encountered with the mash and grain system is how to feed in the proportions that will provide the correct protein intake of approximately 16 percent. Production problems can frequently be traced to a low total intake of protein when a mash and scratch system is being used.

The protein content of most scratch grains is approximately 9 to 10 percent. If, for example, equal proportions of an 18 percent laying mash and scratch grains are fed, the total protein intake will be approximately 13.5 to 14 percent. This will not support optimum egg production—at least during the early part of the laying cycle. Most high-energy complete laying mashes can be safely supplemented with grain at the rate of 1 or 2 pounds per hundred birds per day. Follow the manufacturer's recommendations.

Grain may be fed on top of the mash or in the litter if litter conditions are reasonably dry and clean. When birds scratch for grains it helps to aerate the litter and maintain its condition.

FREE CHOICE OR CAFETERIA STYLE

This method uses a high-protein mash containing from 26 to 40 percent protein, with grain and calcium supplement kept before the birds at all times. The chief advantage of this method is that it enables a producer to use local grain or grain produced on his farm. It is also a simpler system than hand-feeding grain. The mash concentrate is fed in one hopper, and the grains are fed in individual hoppers. The bird actually balances its own nutrient intake, if all goes right.

OTHER FEEDING PROGRAMS

For small-flock situations fresh kitchen scraps, garden products, and even surplus milk can be fed to layers. Feeding these various materials can substantially reduce the amount of purchased grain required. The amounts of these materials should be limited to what the birds can clean up in five or ten minutes. If too much of this sort of material is fed to the birds, the nutritive intake can be diluted to the extent that they don't get the right amount of protein and other nutrients for body maintenance and the production of eggs. Milk should be given to the birds in plastic or enamel containers and not in galvanized containers.

Care should be taken in the choice of table scraps given the birds. Spoiled meat and materials like fruit peelings, onions, and other strong-flavored foods may give a bad taste to the eggs and should not be fed to the birds. Potato peelings can be fed to chickens if they are cooked. Vegetable peelings and green tops of vegetables are good. Green feeds are satisfactory for layers, and those that can be used for laying birds are similar to those recommended for growing birds. When a complete laying mash is supplemented with table scraps and other materials, provide the birds with grit and calcium supplements. Eggshells are a good source of calcium and may be crushed and fed back to the laying flock. A simple or "Natural Laying Hen Diet" for those who want to mix their own ration can be found in Table B (see Appendices).

FEED CONSUMPTION OF LAYERS

Feed consumption varies with the body weight and rate of production (Table 6). A certain amount is used for body maintenance. This amount corresponds with the amount consumed at "0" production. As a rule of thumb to estimate feed consumption at various levels of egg production, add to the amount of feed required for maintenance ("0" production) 1 pound of feed for each 10 percent of egg production.

Birds eat to satisfy their energy requirements, so the feed consumption will vary with environmental temperatures. They will eat more in cold weather. To insure optimum feed consumption, provide 4 linear inches of feeder space per layer.

CALCIUM SUPPLEMENTS AND GRIT

Calcium is one of the most important minerals needed in the layers' diet. Hens need calcium for eggshell formation. About 10 percent of the total weight of the egg is shell, and the shell is almost 100 percent calcium carbonate. Hens in heavy production require calcium in relatively large quantities.

Most of the high-energy complete laying mashes today contain 3 to 3½ percent calcium depending upon the time of year. Under most circumstances, this appears adequate to produce eggs with sound shells.

When laying birds get well into their laying cycle and shell quality begins to deteriorate, or during periods of extremely hot weather when shell quality may

TABLE 6
FEED CONSUMPTION PER DAY PER 100 LAYERS
AND FEED PER DOZEN EGGS
AT VARIOUS LEVELS OF PRODUCTION

PERCENT PRODUC-TION	FEED FOR LEGHORN-TYPE BIRD—4 POUNDS Pounds		FEED FOR BROWN-EGG BIRD—5 POUNDS Pounds		FEED FOR BROWN-EGG BIRD—6 POUNDS Pounds	
	Per Day	Per Doz.	Per Day	Per Doz.	Per Day	Per Doz.
0	15.8		18.6		21.2	
10	16.7	20.1	19.6	23.6	22.1	26.6
20	17.6	10.5	20.4	12.2	22.9	13.7
30	18.4	7.4	21.3	8.5	23.9	9.6
40	19.3	5.8	22.2	6.7	24.7	7.4
50	20.2	4.8	23.1	5.5	25.6	6.1
60	21.1	4.2	24.0	4.8	26.5	5.3
70	22.0	3.8	24.9	4.3	27.5	4.7
80	22.9	3.4	25.8	3.9	28.1	4.2
90	23.8	3.2	26.7	3.6	29.2	3.9

Based on Card, L.E., and Nesheim, *Poultry Production,* Twelfth Edition.

suffer, additional calcium seems to be needed. In these cases, producers commonly add extra calcium to the diet in the form of oyster shells. Some types of birds, particularly the light-laying strains of birds that have relatively small calcium reserves, need calcium supplementation earlier in the laying cycle than some of the heavier types of birds. When to feed supplemental calcium and when not to is still somewhat of a controversial subject. Some feel that the most satisfactory way of meeting the variable individual requirements of layers is to supplement the calcium in the laying mash. This is done by feeding separately hen-size oyster shell or calcite crystals, both of which are soluble.

For those feeding systems other than the complete laying mash system, a supplemental feeding of calcium is recommended.

In addition to a source of calcium a hard, insoluble grit should be fed in many instances. Insoluble grit, such as granite grit, maintains its sharp edges in the digestive system and in essence takes the place of teeth. It is often thought that it is not necessary to feed grit to layers fed an all-mash type of diet. It is assumed that everything the birds consume in the laying diet is already finely ground and, therefore, needs little or no further grinding. Where birds are kept in cages, this is probably true.

Birds housed in a floor management system frequently eat feathers, litter, and nesting material, however. Under these circumstances grit, if available, enables the gizzard to grind the fibrous material so that it can be moved on through the digestive system. Any feeding program other than the all-mash system of feeding should incorporate grit so coarse grains can be utilized by the digestive system.

There are a number of types of grit and shell hoppers available on the market. These are usually divided into two compartments so both grit and the source of calcium can be fed from the same hopper. It is a relatively simple project to construct a hopper that will serve equally well. In those situations where grit and calcium material is needed, the hopper should be kept full at all times.

WATERING

Birds need plenty of clean, cool water at all times if they are to produce well. Water makes up a large portion of the hen's body and is a major constituent of the egg. Water helps to soften the feed and aids in its digestion, absorption, and assimilation. If hens are deprived of water for only a short time, egg production will suffer. Dirty water and watering equipment may discourage water consumption and is a potential source of disease infection. The water supply should never be permitted to freeze for an extended period. If the house temperature cannot be maintained at a level to prevent freezing, an electrical immersion heating unit can be installed in the drinking fountain. Heating units should be thermostatically controlled.

Laying birds normally drink in pounds about twice as much as the feed they consume. During hot weather they will consume substantially more than this amount. One inch of the trough-type waterer per bird should be provided, or one round waterer per one hundred birds. One pan-type waterer is usually adequate for the small family flock. Waterers should be cleaned and filled with fresh water daily. The water consumption of layers is shown in Table 7.

FACILITIES AND MANAGEMENT

In well-insulated, mechanically ventilated houses Leghorn-type layers can be housed at the rate of 1¼ to 1½ square feet per bird. Heavier brown-egg birds should be given 1½ to 2 square feet per bird. In those houses where gravity ventilation is used, that is, ventilation through the windows or slots, slightly more

TABLE 7
WATER CONSUMPTION OF LAYERS
BASED ON ENVIRONMENTAL TEMPERATURE

TEMPERATURE	GALLONS PER 100 LAYERS PER DAY
20–40° F.	4.2– 5.0
41–60° F.	5.0–5.8
61–80° F.	5.8–7.0
81–100° F.	7.0–11.6

Source: *New England Poultry Management and Business Analysis Manual,*
    Bulletin 566 (Revised).

floor space should be given. The normal recommendation is a minimum of 1½ to 2 square feet for Leghorns and 2 to 2½ square feet for brown-egg birds.

Actually, in a well-insulated house cold-weather ventilation is somewhat simplified if there are more birds, since they supply more heat. When feeders and waterers are located over a dropping pit, the floor area per bird can be reduced below the amounts previously mentioned. However, with each additional reduction in floor space per bird, additional feeding, watering, nesting, and roosting space must be provided. The birds should not have to travel more than 15 feet to reach either feed or water.

NESTS

Provide at least one individual nest or 1 square foot of community nest space for each four hens. Nesting material should be replenished as needed and be kept clean and dry. This is very important in preventing dirty, cracked, and broken eggs. The nest should be a minimum of 2 feet from the floor or litter. If floor eggs are a problem at the onset of production, it is advisable to place the nests on the floor and gradually raise them as the birds become accustomed to using them. One of the common problems with small flocks is egg eating. When the nests are poorly constructed or without litter, or the eggs are not gathered frequently enough, breakage occurs, and this habit is much more likely to get started.

EGG GATHERING

Eggs should be gathered two or three times a day. Frequent gathering reduces the number of dirty eggs and improves the egg quality. It also reduces the possibility of birds developing the habit of egg eating. Often it is difficult for small flock owners to gather the eggs as frequently as is desirable. Having plenty of nests, and keeping them clean and with plenty of nesting material,will help prevent broken or dirty eggs. When the flock owner has to be away during the day and the eggs will not be collected, it is possible to have the lights come on earlier in the morning, so that part of the eggs will be laid earlier and can be gathered before leaving for the day. Once egg eating commences it is a difficult habit to stop. Probably one of the best methods of preventing it is to debeak the birds and to increase the frequency of egg collection. Eggs not gathered during the day are also subject to freezing in cold weather.

LIGHTING

The use of artificial light in the poultry house is not intended to give the hens more time to eat. It is the stimulation of the light itself that makes them lay more eggs. The light stimulates the pituitary gland through the eye. This gland, in turn, secretes hormones that stimulate the ovary of the hen to lay eggs.

Laying birds need about fourteen hours of light daily. In the northern part of the United States, darkness exceeds the amount of daylight during many of the fall and winter months. Beginning on August 15th, birds in windowless houses need to receive extra light to give them a fourteen- to fifteen-hour day. During the months of November, December, and January, when the days are shortest, they will need five hours of supplemental light. The lights may be turned on in the morning or in the evening. A combination of morning and evening lights may be used, as long as the birds have at least fourteen hours of light each day.

With small flocks it is a task to turn the lights off each night, or to get up early enough to turn them on in the morning. The best way to solve this problem is to use a time clock which automatically turns the lights off and on at the desired time. This generally gives satisfactory results. As soon as the natural day length reaches fourteen hours again, the lights can be discontinued. Layers housed in controlled-environment or light-tight houses will need to receive a constant fifteen or sixteen hours of light.

The lights should be ordinary 40- to 60-watt incandescent bulbs. Use a reflector to get the most efficient use of the light wattage—this will usually permit the use of lower-wattage bulbs. One light fixture should be installed for each 200 square feet of floor area or less. A distance of 10 feet between lights generally gives good distribution of light for the birds. A rough rule of thumb for light intensity is 1 footcandle at the feeder level. One bulb watt per 4 square feet of floor space will usually provide 1 footcandle if the bulb is 7 to 8 feet from the litter and has a reflector. During the first week or two of the lighting period, it may be necessary to catch and place some of the birds on the perches if they do not get on the roost of their own accord when it's time for lights out.

VENTILATION

During cold weather all of the ventilation should be from one side of the poultry house, preferably the south. Windows that slide down from the top or tip in from the top are best for winter ventilation, in the absence of fan ventilation. The windows should be adjusted according to the weather. When the house tends to be stuffy and the ammonia fumes are strong, the house needs more ventilation. The house should never be closed tight, even on cold nights. It is always best to leave at least some of the windows slightly cracked.

In the warm summer months or in warm climates, ventilation will help cool the house. If you can open up both sides of the house, it will increase the air movement, help keep the birds cool, and maintain better egg production and overall performance.

RODENT AND BIRD CONTROL

Rats and mice like chicken feed. If there is a place for them to hide in the chicken house, they will do it. They enjoy living in piles of refuse, the double walls of buildings, old stone foundations, in dropping pits, under deep litter or under the floors, and in many other hiding places. Good housekeeping will help to reduce the problem with rats and mice.

Anticoagulant rat poisons and other types of poisons have been very helpful in getting rid of rat problems. For anticoagulant-type poisons to be effective, rodents must consume it for several days. If birds are housed in an old building that has plenty of hiding places, it may be best to set up a permanent bait station or stations, and keep bait in them at all times.

It is desirable to screen wild birds out of the chicken house, if possible. Wild birds are vectors of some of the common diseases and parasites of poultry.

CULLING AND SELECTION

Good production-bred chickens will lay well for approximately a year if they are well managed. Culling healthy birds that go out of production for just a short time is not advisable, unless a chicken dinner is in order. After the birds have laid for nearly a year, the nonlayers may be culled out, dressed, and used at home or sold.

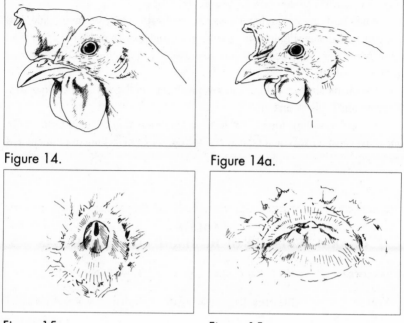

Figure 14.

Figure 14a.

Figure 15.

Figure 15a.

Usually the entire flock should be replaced after twelve to fourteen months of production. On occasion it is necessary—and advisable—to sell hens that are still laying at a reasonable rate, to make room for the new flock of pullets.

In culling, there are two distinct things that you must learn. One is how to distinguish between a layer and a nonlayer. The other is how to determine with a reasonable degree of accuracy how long the bird has been either in or out of production. After you have acquired the ability to do these two things, you can do a reasonably good job of culling a flock of layers.

With the beginning and termination of laying, changes take place in the head, the abdomen, and the vent. To determine whether or not a layer is in production, all three of the body parts should be carefully examined to make an accurate judgment. The examination of one of these parts alone will not provide the whole answer.

When birds are in laying condition, their combs and wattles are enlarged, bright red, and waxy in appearance (Figure 14). The vent is large and moist (Figure 15), and the pubic or pin bones are wide apart to physically permit the egg to be laid. The spread of the pubic bones of the laying bird will be about three fingers wide. The abdomen will be soft and pliable.

When birds are not laying, the combs and wattles may be small, pale, and scaly in appearance (Figure 14a), the vent dry and puckered (Figure 15a). The pubic bones may be close together (one or two fingers width) and the abdomen hard and shallow.

The characteristics of high and low producers and layers and nonlayers are presented in Tables 8 and 8a.

To determine how long a bird has been in production or out of production, one must consider two factors, namely, the degree of bleaching of pigment and the molt.

TABLE 8
CHARACTERISTICS OF HIGH AND LOW PRODUCERS

Character	High Producers	Low Producers
Vent	Bleached, large, oval, moist	Yellow, dry, round, puckered
Eye Ring	Bleached	Yellow-tinted
Beak	Bleached or bleaching	Yellow or growing yellow
Shanks	Pale yellow to white, thin, flat	Yellowish, round, full
Head	Clean-cut, bright red, balanced	Coarse or overrefined, dull, long, flat
Eyes	Prominent, bright, sparkling	Sunken, listless
Face	Clean-cut, lean, free from yellow color and feathers	Sunken or beefy, full, yellowish, feathered
Body	Deep	Shallow
Back	Wide; width carried out to pubic bones	Narrow, tapering, pinched
Plumage	Worn, dry, soiled	Smooth, glossy, unsoiled
Molt	Late molter	Early molter
Carriage	Active and alert	Lazy and listless

TABLE 8a
CHARACTERISTICS OF LAYERS AND NONLAYERS

CHARACTER	LAYING HEN	NONLAYING HEN
Comb	Large, red, waxy, full	Small, pale, scaly, shrunken
Wattles	Large, prominent	Small, contracted
Vent	Large, moist	Dry, puckered
Abdomen	Full, soft, velvety, pliable	Shallow or full of hard fat
Pubic bones	Flexible, wide open	Stiff, close together

Source: *Culling for High Egg Production,* Vermont Agricultural Extension Service Circular 115RU.

PIGMENTATION

The yellow pigmentation found in all yellow-skinned breeds and varieties of chickens is a pigment called xanthophyl. Yellow corn, a major constituent of the poultry diet, is the principal source of this pigment. If a pullet has been properly fed when it comes into the laying house, she will carry a considerable amount of pigmentation in all parts of her body. When she commences to lay, the pigment present in the skin and in other parts of the body is gradually lost. The reason for this loss is that the pigment is diverted to the yolks, giving them their yellow color. As long as the bird is in production, it continues to lose this yellow color from the various parts of its body. Hens that are thoroughly bleached out are usually high producers. When the bird stops production, the pigment reappears in the various parts of the body. The return of pigmentation is one of the first clues that a bird is out of production.

There is a very definite order of bleaching of the body parts as the birds commence to lay—first the vent, then the eye ring, the ear lobe, the beak, the shanks, and the feet. The order of bleaching and the approximate time required to bleach the various parts of the body are given in Table 9.

VENT

The first noticeable change occurs in the skin around the edges of the vent. Within a few days after production commences, the yellow color around the vent disappears.

TABLE 9
ORDER OF BLEACHING AND TIME REQUIRED
WHEN BIRDS ARE IN CONTINUOUS PRODUCTION

Body Part	Time Required
Vent	Few Days
Eye Ring	2 to 3 Weeks
Ear Lobe	3 to 4 Weeks
Beak	6 Weeks
Shanks	2 to 5 Months

A yellow vent indicates that a bird is not laying. A whitish, pinkish, or bluish white vent indicates that a hen is laying.

THE EYE RING

The eye ring starts to bleach soon after the vent. In two to three weeks after the onset of production, the eye ring usually loses its yellow pigment. By that time, the bird has probably laid ten to twelve eggs.

THE BEAK

The yellow color leaves the beak next. It leaves the base of the beak first, and the fading continues toward the tip. It takes approximately six weeks of continuous production for complete bleaching. By that time usually thirty to forty eggs have been laid.

The lower beak bleaches more rapidly than the upper beak. Fading of color can be seen readily in the lower beak. When the dark pigment "horn" appears on the upper beak, the lower beak can be used as a basis for judging the degree of pigmentation loss. Barred Plymouth Rocks and Rhode Island Reds commonly carry this dark pigment.

THE SHANKS

The bleaching of the shanks occurs last and is an indicator of long-term production. Bleaching of the shanks takes from four to five months of continuous production. She must lay from 120 to 140 eggs to be completely bleached. It is not possible to pick a bird as a layer or nonlayer on the basis of bleached or yellow legs unless she has been in production for four or five months.

When the bird stops laying, pigmentation returns to the different parts of the body in the same order that it bleached out.

By observing the degree of pigmentation, one can tell rather accurately how well a bird is laying. It should be taken into consideration that the feeding program, breed, and flock health are factors that can also affect pigmentation.

THE MOLT

Once each year birds renew their plumage. This process of replacing old feathers with new ones is called molt. Hens usually go through their annual molt in the late summer, the fall, or early winter months. Factors that determine the time of the molt are (1) time of the year the bird was hatched, (2) the individual bird or breeding, and (3) management stresses to which the bird is exposed. Just as bleaching is most helpful in determining layers from nonlayers during the first eight or nine months of production, the molt is most useful during the last several months of production.

When a bird starts her molt she goes out of production and, normally, will not come back into production until shortly before the molt is completed or just after it is completed. The pattern depends, to a certain extent, upon the type of management and feeding program to which she is exposed. Some laying birds start their molt early in the fall after eight or nine months of production and are called early molters. Others lay for twelve or fifteen months before they molt and stop production. In fact, some of our modern-day strains of layers have to be force-molted to get them out of production. This procedure provides them with an opportunity to rest, renew their feathers, and return to production. Hens that complete at least twelve months of production before molting are referred to as late molters and are the most desirable birds to have in the flock.

When using the molt to cull or select hens, the primary (flight) feathers of the wing are used. There are usually ten of these feathers, arranged in a group, extending from the short axial feather in the center of the wing to the tip.

The fact that the primary feathers are dropped and renewed in a definite order makes the use of the molt a fairly accurate means of estimating how long a bird has been out of production and how long it will take her to go through her complete molt. It is also possible to determine whether she is a slow molter or a rapid molter by observing the way in which she has dropped and is renewing the primary feathers. A typical slow-molting bird will drop one primary at a time over an extended period, whereas a rapid-molting hen will drop more than one at a time and drop them more frequently. In some of the better strains of laying birds, the hens will drop a group of several primaries and then, very shortly after, drop more and thus go through the complete molt in a relatively short period of time and be back in production. Approximately six weeks are required to grow a primary feather.

Usually the slow molters stop production before or at the time they start their wing molt. They very seldom lay through a molt. Thus, the slow molter that drops one primary at a time will be out of production for many weeks and would be a likely candidate to be culled from the flock. Many of the rapid-molting birds will lay for a period of time after they start to renew their primary feathers. Some of the good laying strains of birds will renew half or more of their primaries before they finally stop laying. Figure 16 shows the different stages of molt in fast- and slow-molting birds.

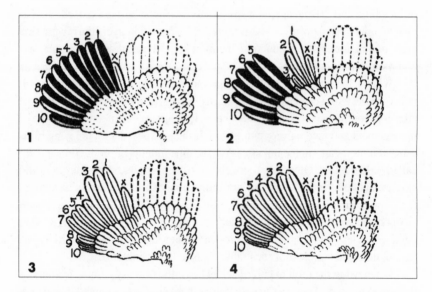

Figure 16. Wings during different stages of molt.

(1) shows the 10 old primary feathers (black), and the secondary feathers (broken outline), separated by the axial feather (x). **(2)** shows a slow molter at six weeks of molt, with one fully grown primary and feathers 2, 3, and 4 developing at two-week intervals. In contrast, **(3)**, a fast molter, has all new feathers. Feathers 1 to 3 were dropped first (now fully developed); feathers 4 to 7 were dropped next (now four weeks old); and feathers 8 to 10 were dropped last (now two weeks old). Two weeks later **(4)**, feathers 1 to 7 are fully grown. Fast molt took 10 weeks, compared to 24 weeks for slow molt. (South Dakota Extension Service)

FORCE MOLTING

Toward the end of the laying cycle, egg production is at a relatively low level. Interior and exterior egg quality are poorer, particularly if the tail end of the cycle coincides with hot weather. At this point some producers force molt or recycle the flock if it has been a good flock.

Force molting gives the layers a rest for about eight weeks. After the rest, egg production increases but not to the level of the pullet year. It may reach 88 to 90 percent of the pullet-year production level. Egg quality, both exterior and interior, recovers substantially.

Recycled layers usually lay profitably for only six to nine months. Egg size is large, feed consumption and mortality is higher, and the overall egg quality is lower than during the pullet year.

The big reason for force molting is to beat the high cost of pullet depreciation which may be 8 to 10 cents per dozen depending upon the pullet cost and fowl or salvage value. If force-molted birds are to be profitable to the commercial producer, there must be a market for the extra-large eggs. The profitability of the

---

FORCE-MOLT PROCEDURE

Day 1 Decrease lights to 8 hours per day in light-tight houses or no-artificial-light in window houses. Remove all feed and water.**

Day 3 Provide water.

Day 8 Provide growing mash—40% normal consumption level.

Day 22 Restore lights to pre-molt level and full feed laying ration.

An **alternative method, particularly during periods of hot weather, is as follows:

Day 1 Decrease lights to 8 hours per day in light-tight houses or no-artificial-light in window houses. Remove feed but provide water free choice.

Day 8 Provide growing mash—40% normal consumption level.

Day 22 Restore lights to pre-molt level and full feed laying ration.

From *New England Poultry Management and Business Analysis Manual*, Bulletin 566 (Revised).

practice is still under investigation. Most producers feel that it is better to replace the flock at the end of the first laying cycle.

BROODINESS

Poultry geneticists have done a great deal to eliminate the natural tendency for laying birds to become broody. There will, however, still be an occasional bird in floor management systems that will become broody. If the bird is permitted to stay in the nest, she will remain broody and will not eat and drink properly during this time, and a loss in body weight and egg production will result. If broody hens are removed from the nest as soon as they are discovered, the loss of egg production can be reduced substantially.

The best procedure for breaking up broody birds is to place them in a broody coop—constructed with a wire or slatted bottom to discourage setting and inactivity. The broody coops are usually suspended from the ceiling in the pen or outside the pen, and the birds are provided with plenty of feed and water. Birds exposed to this type of treatment can usually be put back with the flock in four or five days.

FLOCK HEALTH

Flocks that are well managed and fed a well-balanced diet do not experience so many disease problems. If one bird dies now and then it is probably nothing to become alarmed about. If a number of birds become sick, respiratory symptoms appear, or they look droopy, go off feed, or stop laying, it is then time to find out why.

If a disease outbreak occurs, the cause should be determined as soon as possible. First signs are often a change in feed and water consumption. If you do not recognize the disease or parasite, take or send some of the live birds showing symptoms to the nearest poultry diagnostic laboratory. Poultry diagnosticians prefer live birds showing symptoms. Dead birds sent to the laboratory should be kept cool so that they do not arrive at the laboratory in a decomposed condition. It is helpful to poultry diagnosticians to know the flock history. Relate the symptoms you have observed in the flock, the number of birds affected, the number of birds that have died, source and size of the flock, feeding program, age of the flock, and any other information that you think will be helpful to them.

To control disease on the farm, incinerate dead birds or put them into a disposal pit and remove all obviously sick chickens from the flock. It is sometimes best to kill them and dispose of the carcasses. If, however, the birds are to be treated, put them in a separate pen or hospital pen—as far away from the other birds as possible.

The old adage, "an ounce of prevention is worth a pound of cure," certainly applies to poultry flock health. If you are successful in buying or raising good healthy pullets, they are properly vaccinated, placed in cleaned and disinfected quarters, given enough feeder, water, and floor space, and are fed a well-balanced diet, you are well over the hurdle with regard to disease or production problems.

Chickens are creatures of habit; any sudden changes can cause them stress. This is true of layers in peak production and those nearing the end of their laying cycle. Any change in feed or management should be done gradually to avoid stress problems.

Birds should be vaccinated for such diseases as infectious bronchitis, Newcastle disease, fowl cholera, laryngotracheitis, and fowlpox, if these are a problem in your area. A good laying-flock manager checks for internal and external parasites and immediately commences control measures as indicated. Screening windows, doors, and other points of entry for wild birds will help to prevent some disease and parasite problems.

The immediate disposal of dead birds by burying or incinerating is recommended. Check local and state ordinances concerning the disposal of dead birds in your area. Some types of incinerators are not approved in certain states.

Cannibalism, feather pulling, and egg eating are habits that are sometimes hard to stop once they get started. Birds commence these vices usually as a result of faulty management or environment. This trait is more common in some breeds of chickens than in others. Different types of feed may also cause these problems. Crumbles and pellets can bring on the picking problem when fed instead of mash. The birds fulfill their intake needs more quickly on pellets and have time to get into mischief. Proper debeaking is usually the answer when those problems begin. Actually, the birds should be properly debeaked by sixteen weeks of age so that it doesn't have to be done after the birds come into production. Debeaking after production commences can cause a production slump.

Frozen combs are the small flock hazard in cold climates. Some breeds and varieties, such as the Single Comb White Leghorn, and the males of many breeds have large combs. These are quite easily frozen in cold housing and can cause lowered egg production and a reduction of egg fertility in mated flocks. When severely frozen, the top of the comb becomes discolored and eventually may slough off. Naturally there are sore heads, and feed consumption may be discouraged by using grills or reels on feeding and watering equipment (see Chapter 2).

This problem can be avoided by dubbing the birds. The best time to do this is at the hatchery or the farm at day of age. It is done with a pair of manicure scissors. The comb is cut off close to the head, giving a closely cropped comb at maturity, one which is less likely to be frozen or injured. Clean equipment in a mild chlorine solution between birds: ten drops of chlorine bleach or Clorox per quart of water. White Leghorns that are to be put in laying cages are frequently dubbed to avoid injury which may occur from the cage wires. Dubbing can be done to older birds, at which time both the comb and wattles are cut with dull shears. Considerable bleeding occurs in older birds.

RECORDS

Recordkeeping is not only interesting but necessary if you want to know what the flock is doing and the costs and returns involved. Records need not be complicated but merely contain information on daily egg production, mortality, culling, feed consumption, and the quantity and value of poultry and eggs eaten and sold.

Records may be kept on a calendar or on a pen record obtainable from most feed companies. You can make your own pen record. At the end of the month the information can be transferred from the pen record to a permanent record. These records should be analyzed each month and at the end of the year. Records should be filled in each day to be accurate. Without good records it is hard to tell if you should continue in business or buy your eggs and poultry at the store.

COSTS OF EGG PRODUCTION

Budgets for small flocks can vary substantially from the budget information in Tables 10 and 11. Overhead costs, interest, and labor are frequently not considered in the small-flock situation. Bird depreciation may vary considerably. The fowl may be processed and used at home or sold, and would thus be valued at the consumer price instead of the farm price for fowl, which is relatively low. Feed price is another variable. Small-flock owners normally buy their feed in small quantities at a price well over that paid by the commercial producer. Feed prices also will depend upon the feeding program, the geographical area, ingredient supplies, prices, and other factors.

The two budgets are based on the following assumptions. Birds are housed at 22 weeks of age, sold at 73 weeks of age (51 weeks of production).

Annual mortality: 7%

Eggs per hen housed: 253
Feed per 100 birds daily
    Brown-egg birds: 25.2 pounds (90.5 pounds annually)
    White-egg birds: 21.9 pounds (78.5 pounds annually)

TABLE 10
COST PER DOZEN ANALYSIS—WHITE EGGS

ITEM	COST/CENTS
Feed (3.71 pounds) at $10.00 per cwt	37.10
Bird depreciation	12.40
Overhead	2.82
Other	1.12
Interest	2.11
Labor	2.29
TOTAL	57.84

Based on information from the *New England Poultry Management and Business Analysis Manual*, Bulletin 566 (Revised).

TABLE 11
COST PER DOZEN ANALYSIS—BROWN EGGS

ITEM	COST/CENTS
Feed (4.28 pounds) at $10.00 per cwt	42.80
Bird depreciation	11.80
Overhead	3.76
Other	1.45
Interest	2.66
Labor	2.29
TOTAL	64.76

Based on information from the *New England Poultry Management and Business Analysis Manual*, Bulletin 566 (Revised).

CHAPTER 6

The Production of
Fertile Eggs

A ROOSTER is not needed in the laying flock—the hens will lay just as well without him. Some individuals do, however, like to mate their flock of layers merely to produce fertile eggs. Others plan to perpetuate a breed or strain of bird as a hobby, for exhibition purposes, to sell to other poultry fanciers, or for a number of other reasons.

If flocks are mated to produce eggs for hatching, it is good to check on state regulations. Some states require that all hatching-egg flocks be blood-tested for Salmonella Pullorum and Salmonella Typhoid. Blood testing is supervised by employees of the official state agency that administers the National Poultry Improvement Plan in the state. For further information on blood-test requirements, check with your State Department of Agriculture, state Extension Poultryman, or county Extension Agent.

As is the case with other laying flocks, careful feeding and management, good housing, and proper equipment are important if one is to get a good hatch of healthy chicks from the eggs.

Select a source of stock that is healthy and bred to meet the specifications you have set, whether it is feather color, body size, egg production, meat production, or a combination of these characteristics.

MATING

A hatching-egg flock of Leghorn or Leghorn-type birds should contain about six to seven males per one hundred pullets. Heavier breeds, such as the Rhode island Red and the Plymouth Rocks are mated at the rate of eight males per one hundred pullets. The ratio of males to pullets in some of the commercial hatching egg flocks may vary from this ratio substantially, due to anticipated mortality, morbidity, or other special problems with the males.

If you are trying to perpetuate a breed, the breeding stock should be selected to conform to those characteristics desirable for that breed or variety, as outlined in the American Standard of Perfection, issued by the American Poultry Association. It describes the various classes, breeds, and varieties of poultry recognized as standard-bred. Primary breeders select for numerous economic characteristics, including production, production efficiency, egg size, egg quality, livability, and many others.

If yearling hens are to be mated, force them out of production to permit approximately an eight-week rest before they commence laying hatching eggs. The same force-molting procedure is used as outlined in Chapter 5.

The flock should be mated at least two weeks before hatching eggs are to be saved.

FEEDING AND MANAGEMENT OF BREEDERS

In addition to producing eggs efficiently, breeding stock should produce eggs that hatch well and produce vigorous chicks. Breeder flocks should be fed a diet that has additional vitamins and minerals for high fertility and hatchability. For best results, then, the breeding flock should be fed a specialized diet formulated for hatching-egg production. These are known as breeder rations.

The management of a breeding flock is very similar to a market-egg flock. Most of the management recommendations given in the preceding chapter on laying-flock management, therefore, need not be repeated in this chapter. Only the basic differences between breeder-flock and laying-flock management will be mentioned.

One of these basic differences is the management system required. It was mentioned in the preceding chapter that there are several management systems available for laying birds. Among those mentioned were the cage system, the combination slats and litter, the all-slat or the combination wire-and-litter, the all-wire, or the conventional litter-management approach.

Cage management systems for breeding birds are relatively uncommon. This is because of the difficulty in designing cages that allow birds to mate with good results. Where cage systems are used it is usually necessary to artificially inseminate the females. This is not a difficult task where small flocks are involved, but is much more time consuming than the conventional floor-pen matings.

Other management systems have been used with some degree of success, but there are a number of problems involved with systems other than the littered floors. This is particularly true with the heavier birds that appear to be inherently lazy and don't want to lay well in the nest. For example, the all-slat floors or even the slat-and-litter management systems have often produced a high percentage of floor eggs or eggs laid on the slats. This can cause dirty eggs, broken eggs, and the loss of eggs through the slats. It is probably best not to use wire floors for mated flocks, particularly the heavy-breed flocks, because of the incidence of foot and leg problems and the relative difficulty for the birds to mate.

LIGHTING THE BREEDERS

Light is important not only to stimulate egg production in breeding flocks, but also to increase the semen output of the males. Lighting schedules were outlined in Chapter 5, Managing the Laying Flock. Schedules vary with the type of house. Careful attention to lighting should be given.

Hatchability of fertile eggs is affected by many factors. Shell quality is one of these factors. The longer the bird is in production, the poorer the shell quality. Shell quality is affected not only by the length of time the birds have been laying, but also the time of year, temperature, diet, and genetics. Birds that have been force molted and are in a second year of egg production invariably show shell-quality deterioration much more rapidly than during the first laying cycle.

Tremulous or loose air cells also affect hatchability; use care in handling eggs to prevent this. Cracked eggs very seldom hatch, so take care to produce as few cracks as possible and to make sure that cracked eggs are not set in the incubator.

CARE OF HATCHING EGGS

Hatching eggs should be gathered three to four times a day. When temperatures are extremely hot or cold, they should be gathered more frequently. Eggs should not be left in the nest overnight, and they should be cleaned as soon after gathering as possible.

CLEANING

Unless they can be washed properly, they should not be washed but rather dry-cleaned by sanding or buffing. Hand-buffing devices are available for this purpose.

Eggs should be washed at a temperature of 110° to 115° F. in water containing an approved detergent-sanitizer. They should not be immersed in the wash water for more than three minutes. The eggs should be dried and moved to cold storage. Improperly cleaned eggs frequently become contaminated and may explode during incubation.

STORAGE

Store eggs at a temperature of 50° to 60° F. Lower storage temperatures may reduce hatchability. The temperature of the average household refrigerator runs below that recommended for storing hatching eggs. For best results, keep the relative humidity at 75 percent.

Eggs that are cracked, have thin shells, shells with ridges, or that are excessively dirty or abnormal in size or shape should not be kept for hatching. Excessively large or small eggs are often infertile or won't hatch, and should not be set in the incubator.

For best results, hatching eggs should not be stored for more than ten days to two weeks before they are set. They should be stored in filler flats or egg cartons with the small end of the egg faced down. Eggs that are to be held for several days prior to hatching should be held in a slanted position (approximately 35°) and turned at least twice each day. One simple method of turning eggs is to prop one end of the egg carton or case up on an object at an angle of about 35°. Then just shift ends with the container at least twice each day.

INCUBATION MACHINES AND OPERATING TEMPERATURES

There are two types of incubators in use today. These are the forced-draft and the still-air machines. The one most frequently used in commercial hatcheries is the forced-draft machine. This type of incubator has fans that force the air through the machine and around the eggs. For most types of eggs the incubator temperature of the forced-draft machine is set at 99½° to 99¾° F.

Most of the still-air incubators in use today are quite small. They are made to hold from one to one hundred or more eggs. Still-air machines do not have fans. They depend upon gravity ventilation through vents on the top and bottom of the machine. The operating temperature of the still-air machine is higher than for the forced-draft incubator. It ranges from 101½° to 102¾° F., depending on the type of egg being incubated. Actually, good results may be obtained by using a constant 102° F. without hurting the embryos, as long as the temperature does not stay at these extremes.

VENTILATION

Ventilation is very important in the incubator. Without it the embryos will suffocate. For machines with adjustable vent opening, the openings are usually cracked open at the start and are gradually opened to permit more ventilation toward the end of the incubation period. Follow the manufacturer's directions, or if using a homemade machine, experiment until the best results are obtained. The room where the incubator is located should be ventilated, yet not drafty.

HUMIDITY

Humidity in the incubator varies from 83° to 88° F. (wet bulb thermometer) depending on the type of eggs. This level of humidity is maintained until the last three or four days before hatching, and then is increased to 90° to 95° F. (wet bulb).

The humidity in small incubators is provided by moisture evaporation pans. The pans have to be kept full at all times. When adding water, add warm water (110° F.) so as not to reduce the incubator temperature for an extended period of time. If the humidity in the machine needs to be elevated, a large sponge placed in the evaporation pan will help to increase the moisture level. More pan space also helps.

During the incubation period (nineteen days), eggs should lose about 11 percent of their original weight through evaporation. A loss of more than this amount is detrimental. The amount of evaporation is controlled by proper humidity. The relative humidity level for chicken eggs should be 50 to 55 percent for the first eighteen days (83° to 87° F. wet bulb reading) and 65 percent (89° to 90° F. wet bulb reading) during the last three days. If humidity is too low, close the vents part way, or add to the moisture evaporation surface. If humidity is too high, permit more ventilation.

Figure 17 shows the approximate size of the air cell when evaporation is normal. By candling the egg at various stages of incubation, this can be used as a rough guide to control humidity in the absence of a wet bulb thermometer. The candling procedure and candling light are covered in the section on candling, ahead in this chapter.

TURNING THE EGGS

While eggs are incubating they should be turned at least four times a day to prevent the embryo from sticking to the shell membrane. Large machines may include automatic or mechanical turning devices. Small incubators sometimes

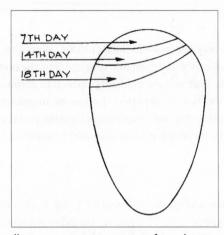

Figure 17. Air-cell size at various stages of incubation.

have turning devices, but usually the eggs have to be turned by hand. If the eggs are not turned during the night, they should be turned late in the evening and early in the morning. If eggs are set on their ends, they should be slanted at about 30°. If they lie flat in the tray, they should be turned from one side to the opposite—in other words, 180°. One method of making sure that all eggs are turned is to put an "X" on one side of the egg and an "O" on the other, in pencil. When turning the eggs make sure all the "X's" or "O's" are on top. Discontinue turning after the eighteenth day for chicken eggs, or three days prior to hatching for eggs of all other species.

CANDLING

Eggs are normally candled three days prior to hatching and the infertiles are removed. Candling is done in a dark room using a special light. Figure 18 shows a candling light that can easily be made at home. If the eggs are infertile, they will appear clear before the candling light. The fertile eggs will permit light only through the large, or air-cell, end of the egg. The rest of the egg will be black or very dark in color. Eggs may be candled earlier in the incubation period: at seventy-two hours the early fertiles will have the typical blood vessel formation, looking much like a spider.

After the eggs hatch, leave the young birds in the incubator for approximately twelve hours until they are dried and fluffy before removing them. The young can survive without food and water for seventy-two hours, but the sooner they are put on feed and water, the better.

Table 12 contains incubation trouble-shooting information.

TABLE 12
INCUBATION TROUBLE CHART

Symptom	Probable Cause	Suggested Remedies
Eggs Clear No blood ring or embryo growth	1. Improper mating. 2. Eggs too old. 3. Brooding hens too thin. 4. Too closely confined.	1. 8 to 10 vigorous males per 100 birds. 2. Eggs set within 10 days after date laid. 3. Keep hens in good flesh.
Eggs Candling Clear But showing blood or very small embryo on breaking	1. Incubator temp. too high. 2. Badly chilled eggs. 3. Breeding flock out of condition (frozen combs, chicken pen). 4. Low vitamin ration.	1. Watch incubator temp. 2. Protect eggs against freezing temp. 3. Do not set eggs from birds with frozen combs or with contagious diseases. 4. Feed fish oil and alfalfa.
Dead Germs Embryos dying at from 12-18 days	1. Wrong turning. 2. Lack of ventilation. 3. Faulty rations.	1. Close temp. regulation. 2. Plenty of fresh air in incubator room and good ventilation of machines. 3. Feed yellow corn, milk, alfalfa meal, and fish oil.
Chick Fully Formed But dead without pipping	1. Improper turning. 2. Heredity. 3. Wrong temperature.	1. Turn eggs 4 times daily. 2. Select for high hatchability. 3. Watch incubator temp.
Eggs Pipped Chick dead in shell	1. Low avg. humidity. 2. Low avg. temperature. 3. Excessive high temp. for short period.	1. Keep wet bulb temp. from 85°-90° F. 2. Maintain proper temp. throughout hatch. 3. Guard against temp. surge.
Sticky Chicks Shell sticking to chick	1. Eggs dried down too much. 2. Low humidity at hatching time.	1. Carry wet bulb temp. at 85° F. between hatches. 2. Increase wet bulb reading to 88°-90°when eggs start pipping.

Symptom	Probable Cause	Suggested Remedies
Sticky Chicks Chicks smeared with egg contents	1. Low avg. temperature. 2. Small air cell due to high avg. humidity.	1. Proper operating temp. 2. Increase ventilation and lower humidity.
Rough Navels	1. High temp. or wide temp. variations. 2. Low humidity.	1. Careful operation. 2. Proper humidity.
Chicks Too Small	1. Small eggs. 2. Low humidity. 3. High temperature.	1. Set nothing under 23-oz. eggs. 2. Maintain proper humidity. 3. Watch incubator temp.
Large Soft-Bodied Chicks	1. Low avg. temperature. 2. Poor ventilation.	1. Proper temperature. 2. Adequate ventilation.
Mushy Chicks	1. Navel infection in incubator.	1. Careful fumigation of incubator between hatches.
Short Down on Chicks	1. High temperature. 2. Low humidity.	1. Proper temperature. 2. Careful moisture control.
Hatching Too Early With bloody navels	1. Temperature too high.	1. Proper control of temp.
Draggy Hatch Some chicks early, but hatch slow in finishing	1. Temperature too high.	1. Proper operation.
Delayed Hatch Eggs not starting to pip until 21st day or later	1. Avg. temp. too low.	1. Watch temp., check thermometers.
Crippled Chicks	1. Cross beak—heredity. 2. Missing eye—abnormal. 3. Crooked toes—temp. 4. Wry neck—nutrition?	1. Careful flock culling. 2. Matter of chance. 3. Watch temperature. 4. Not fully known.
Excessively Yellow	1. Too much formaldehyde, fumigation.	1. Follow directions on fumigation program.

Source: *Embryology and Biology of Chickens*, University of Vermont.

Table 13 lists the incubation period for eggs of various species.

TABLE 13
INCUBATION PERIODS FOR VARIOUS SPECIES

SPECIES	DAYS
Chicken	21
Duck	28
Muscovy duck	33–37
Turkey	28
Goose	30–32
Guinea	26–28
Pheasant (Ring Neck)	23–24
Mongolian Pheasant	24–25
Ostrich	42
Pigeon	16–20
Japanese Quail	17
Bobwhite Quail	23
Peafowl	28

NATURAL INCUBATION

The first requirement for hatching eggs naturally is a good setting hen. Many of our modern-day chickens have had the broodiness characteristic bred out of them. Rhode Island Reds, Plymouth Rocks, Wyandottes, and Orpingtons are usually good brooders. Don't use a bird with long spurs: she is likely to trample the chicks. Select a calm bird who will be less likely to break eggs in the event that she becomes frightened.

Provide a suitable nesting box, one that is roomy and deep so that she will have plenty of room to change her position, turn the eggs, and be comfortable. Select a quiet spot for the nest, away from dogs or other predators and disturbances. Darken the nest area and provide feed and water close by. The bird should be free of lice or mites before setting.

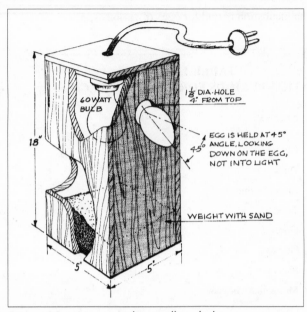

Figure 18. Homemade candling light.

CONSTRUCTING A SMALL INCUBATOR

A small still-air or forced-draft incubator can be constructed quite easily and inexpensively (Figure 19). There are several ways to make them. If the incubator is to be used for demonstration purposes, the top or one side can be made of glass

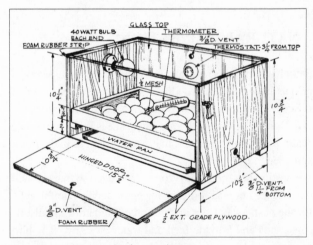

Figure 19. Homemade incubator.

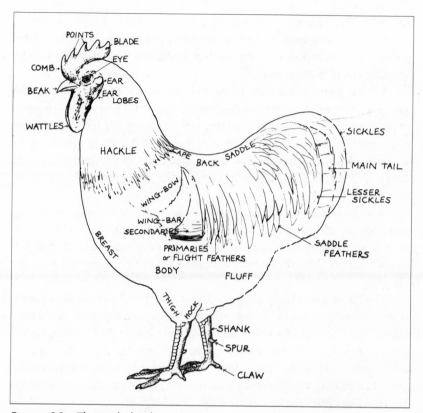

Figure 20. The male bird.

so that the hatching process can be observed. Some excellent still-air incubators have been constructed from Styrofoam coolers. Others have been made from cardboard boxes or fabricated from plywood.

Some commercial concerns offer incubator plans and kits for sale, and of course, some companies also sell incubators. A list of these concerns selling incubators and incubator parts may be found in *Sources of Supplies and Equipment,* in the back of this book.

SEX IDENTIFICATION

Chicks can be purchased straight-run from the hatchery. Straight-run or as-hatched chicks will be 50 percent females and 50 percent males. Small flock owners sometimes order straight-run, dual-purpose chicks and raise the males for broilers and roasters and keep the pullets as layers. Large commercial producers and most small-flock producers prefer sexed chicks, however—either pullets for layers or perhaps males for meat production. Thus there is a need for sex differentiation at the hatchery.

Several methods are used in the industry to sex day-old chicks. The Japanese, or vent-sexing, method has been used for many years. The procedure requires an examination of the rudimentary copulatory organs. An efficient sexer must have good eyesight, be well trained, and keep in practice to be accurate. There is great variation in the appearance of the male and female organs, making it a challenge for the inexperienced to accurately determine the sex of day-old chicks.

Another method often used involves sex-linked inheritance. Sex-linked characters are transmitted from the female of a given mating to her sons but not her daughters. When sex-linked genes produce recognizable characteristics in newly hatched chicks, these characters can be utilized to identify the sex of the offspring.

The Rhode Island Red male crossed with the Barred Plymouth Rock female is an excellent example of sex linkage and its use in identifying the sex of baby chicks. The female chicks will be nonbarred, and the male chicks will be barred. The barring will not show on the down of the newly hatched chicks but will show up as a white spot on the back of the head. The females will not have the spot and when mature will be predominantly black, while the males will exhibit barring. The chicks can be sorted easily.

A second example of sex linkage is rate of feathering. Slow feathering is dominant to fast feathering, so the cross is made between a rapid-feathering male and slow-feathering female. At hatching, the rapid-feathering female chicks from this cross will show well-developed primary and secondary wing feathers, while the slow-feathering males will have primaries and secondaries that are much shorter.

THE AVIAN EGG

The avian egg is one of nature's marvels. We tend to think of it as a source of food, and certainly the egg is important as a food. In fact, it is probably the most nearly complete food known to man. The Poultry and Egg National Board refers to it as "the incredible edible egg."

Nature, however, intended the egg to be a means of reproduction. It is actually a single reproductive cell surrounded by all the nutrients needed by the developing embryo. In fact, it provides enough residual nourishment for the baby chick to maintain itself for seventy-two hours after hatching.

The egg (ovum) remains a single cell until fertilized by the single cell (nucleus) of the male sperm. Fertilization occurs about twenty-four hours before the egg is laid. At that point it carries the full complement of chromosomes and genes. The completed single cell (zygote) rapidly divides into two cells, four, eight, sixteen, etc., after fertilization. The hen's body temperature is 105° to 107° F., so cell division continues; by the time the egg is laid, several thousand cell divisions have taken place. When the egg is laid and cools, cell division stops, to be resumed when exposed to the right conditions of moisture and heat. In nature or where permitted, the female provides these conditions when she sets on the eggs. It is done artificially with an incubator.

Most people, young or old, are fascinated by chicks hatching from eggs. Several projects using the egg as a study tool have been designed to help provide an understanding of some of the principles of reproduction. Before discussing these, it is important to look at the parts of the avian egg (Figure 34) and their role in the process of embryo development, as well as the reproductive system of, particularly, the female, egg formation, and how fertilization takes place.

The hard protective covering of the egg (shell) is composed primarily of calcium carbonate. The shell contains approximately seven thousand pores. Those at the large end of the egg are larger and more numerous. The pores permit the exchange of gases: carbon dioxide and moisture are given off through the shell pores, and atmospheric gases such as oxygen are taken in.

Two tough membranes are attached to the shell. They are the outer and inner shell membranes, and they prevent rapid evaporation of liquid from the egg and also prevent bacterial contamination.

As the egg cools after being laid, the egg contents, which practically fill the shell at this point, contract, bringing in atmospheric air through the pores at the large end of the egg. The air cell is thus formed, providing the first air for breathing when the chick begins to pip its way out of the shell.

The *albumen* provides a liquid medium for the developing embryo. It also contains a large amount of protein necessary for embryonic growth.

In a fresh egg, two white twisted cords are attached to the yolk on one end and into the outer layer of thick albumen on the other end. These are called *chalazae*. The other ends of the chalazae are attached to the outer layer of thick albumen. The chalazae hold the yolk centered in the egg and keep the yolk from adhering to the shell membranes and damaging the embryo.

The yolk contains large amounts of protein, carbohydrates, fats, vitamins, and minerals, all essential for normal growth. These substances combine with oxygen taken in through the shell pores to provide an abundant source of metabolic energy for the developing embryo. By-products of these metabolic processes are carbon dioxide and water. The carbon dioxide is given off through the shell. The water is used by the embryo to replace that lost through evaporation. Calcium absorbed from the shell and the yolk is used for the development of the skeletal system.

THE REPRODUCTIVE SYSTEM AND FERTILIZATION

The reproductive system of the egg has two parts: the *ovary* and the *oviduct* . Unlike most female animals, the hen has only one functioning ovary: the left ovary.

THE OVARY

The ovary is a cluster of developing yolks attached to the hen's backbone. It is located about midway of the back. The ovary contains several thousand immature ovum when the female chick is hatched. As she reaches maturity, the ova start to develop a few at a time.

THE OVIDUCT

The oviduct is a tubelike structure lying along the backbone. It is approximately 27 inches long in a mature fowl. When the yolk is completely formed on the ovary, the follicle sac ruptures, releasing it from the ovary. The yolk then enters the oviduct entrance, or *infundibulum* or *funnel*. During the approximately twenty-four hours following ovulation, the yolk moves on through the oviduct and the chalazae, albumen, shell membranes, and shell are formed around the yolk.

The male has two testes located high up in the body cavity along the back. When the male mates with the female, he deposits spermatozoa into the oviduct. The sperm containing the male germ cells travel on through the oviduct and are stored in the infundibulum.

The egg yolk has a tiny white spot on the surface called the *blastodisc*. The blastodisc contains the single female cell. When the yolk enters the infundibulum, a sperm penetrates the blastodisc and fertilizes it. The blastodisc then becomes a *blastoderm*.

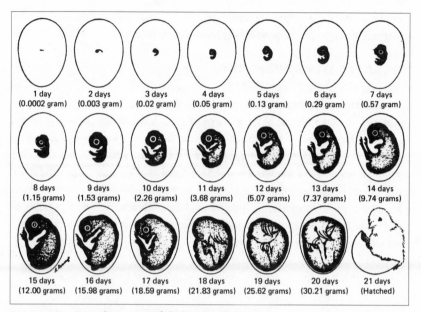

Figure 21. Development of the embryo.

METHODS OF OBSERVING
EMBRYO DEVELOPMENT

A number of methods can be used to observe embryo development during the incubation period. Some of these include candling, the shell window method, and the in vitro method. Egg candling was discussed earlier. It is a very useful

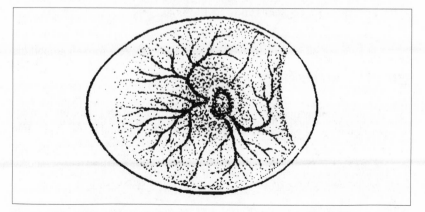

Figure 22. Four -day-old chick embryo seen through the shell.

method to determine if eggs are fertile early in the incubation period, and also for observing embryo development later on in the incubation period.

THE SHELL WINDOW METHOD

As the name implies, the shell window method involves cutting the large end of the egg to allow a view of the embryo. Remove the shell carefully to avoid damaging the embryo. Crack the shell and cut or peel it away, using care not to puncture the inner shell membrane. After the shell is removed, peel away the inner membrane using forceps, tweezers, or scissors. The embryo may be stuck to the inner shell membrane. If so, moisten the membrane with warm water from a small dropper to loosen it, and peel away the membrane. It is now possible to observe the embryo through the window in the shell. If put back in the incubator, the embryo will survive for a few hours.

EQUIPMENT NEEDED

Incubator—with three-day embryos.

Rings—4-inch diameter plastic pipe cut in 2-inch slices.

Plastic wrap—inexpensive or generic is best. Expensive wrap or microwave wrap is not porous enough to permit exchange of gases needed for healthy embryos.

Rubber bands—strong and fairly large; need two per ring.

Scissors—used to trim the wrap.

Receptacle— for eggshells and paper towels.

THE IN VITRO METHOD

The in vitro procedure provides an excellent means of observing embryo development without the interference of the shell. This method makes it possible to observe the early development of the circulatory system with the heart beating outside the body. The eye may also be seen easily after the third day of incubation. As the embryo slowly develops, it is easy to observe the process and day-to-day changes.

Three-day embryos are removed from the shell and placed in plastic pouches or slings. The pouch is made of clear plastic and suspended by a ring. The containers may be removed from the incubator periodically for observation and then placed back in the incubator for further development to occur. With care, many of the in vitro embryos live for several days.

Fertile eggs must be incubated a full seventy-two hours before they are ready to be prepared for the in vitros. Directions for preparation follow.

1. Cut two 1-foot lengths of plastic wrap for each ring to be prepared.
2. Drape one piece of plastic wrap over the ring. Using both hands, place one of the rubber bands around the ring and the plastic wrap. Let it roll off your fingers ½ inch from the bottom of the ring and adjust the wrap to remove the wrinkles.
3. Using the fingers of both hands, slide the wrap and rubber band up to within ½ inch of the top of the ring. This will permit the accumulation of enough wrap to form a pouch or cradle for the embryo inside the ring. Push down gently on the wrap to form the pouch. The bottom of the pouch should extend down in the center to about ½ inch from the bottom of the ring. Adjust if necessary.
4. Remove each three-day embryo from the incubator one at a time as you prepare the in virtro, so as to avoid chillling the embryos. Hold the egg still on its side for twenty-five seconds in the position you would if you were going to break it out into a frying pan for cooking. This enables the embryo to float to the top and helps to prevent the rupture of blood vessels when the egg is broken out into the pouch. Strike the underside of the egg on a hard surface, lower the egg as close to the bottom of the pouch as you can, and gently pry open the shell and deposit the egg contents into the pouch. What you should see on the top of the yolk is a faint circle of blood vessels about the size of a nickel. The heart is deep brown in color and shaped like a comma; it should be beating. If the yolk is clear, turn it over,using your fingers, and look for signs of embryo development.
5. When you have determined that you have a viable embryo, drape the second piece of plastic wrap over the ring. Place the second rubber band over the wrap and around the ring, and adjust to pull the wrap tight. Trim the excess wrap away. Keep the ring on a flat surface while preparing the in vitro. Be careful not to jiggle or tip any more than necessary.
6. Place the in vitros in the incubator as they are completed. In vitros are not turned in the incubator. Several in vitros should be prepared, to insure an adequate number of observations and enable comparison. The purpose of the in vitro is to make embryo development observable. It has to be handled

to be seen, but be careful. Avoid jarring, tipping, or excessive cooling. Two to three minutes out of the incubator should be sufficient for most observations.

It is difficult to predict how many embryos will die and when. Many die at six to eight days, yet others live sixteen to nineteen days.

CHAPTER 7

Meat Bird Production

BIRDS GROWN for meat include broilers, roasters, capons, turkeys, and waterfowl. Turkeys and waterfowl seemed important enough in themselves to warrant separate chapters. Thus, they are covered in Chapters 8 and 9.

Most meat-type birds, and particularly broilers, were once produced as a by-product of chick replacements for egg production. Chicks were purchased as straight run, that is approximately 50 percent cockerels and 50 percent pullets. The cockerels were separated from the pullets at about ten to fourteen weeks of age and marketed as broilers or fryers or carried on to heavier weights as roasters. Males to be used as capons were taken out, or separated from the pullets, at about five or six weeks and surgically caponized and carried through to marketing time. Today these cockerels are destroyed at the hatchery, in most instances, because they are relatively inefficient producers of meat.

For top performance, birds to be grown for meat should be selected from those sources that have outstanding broiler birds. Today's broiler chick is a highly specialized bird. It is not a good egg producer but very efficient as a producer of meat. Broiler chicks are usually crossbreds or hybrids and the parent stock that produces the chicks are usually of White Cornish and White Plymouth Rock breeding. White feathers are preferred in meat birds for the ease of dressing and for better carcass appearance. The Cornish White Rock crosses give chicks with white or predominantly white feathers and also produce chicks with meaty breasts, large legs, and with rapid, efficient growth characteristics. The dressed birds have excellent conformation and market appearance due to the lack of dark pinfeathers.

Breeders are continually improving their stocks. When buying a broiler chick or any chick to be used for meat production, check the growth rate and feed conversion of available stocks in your area. Contact your state Poultry Extension specialist or county agent. It pays to start with the best chicks you can get. Chick cost is a small portion of the total cost of producing a pound of poultry meat.

Table 14 presents information on broiler growth, feed consumption, and feed conversion. Water consumption data is presented in Table 15. The costs

TABLE 14

STANDARDS FOR BROILER GROWTH AND FEED CONSUMPTION

WEEK	LIVE WEIGHT (POUNDS)			GAIN OVER PRECEDING WEEK (MIXED SEXES) POUNDS	FEED PER BROILER FOR THE WEEK (MIXED SEXES) POUNDS	FEED TO DATE (MIXED SEXES) POUNDS	FEED CONVERSION* TO DATE (MIXED SEXES) POUND OF GAIN TO POUND OF FEED
	COCKERELS	PULLETS	MIXED SEXES				
1	.33	.32	.33	.24	.29	.29	.88
2	.85	.79	.82	.49	.61	.90	1.10
3	1.56	1.41	1.48	.67	.99	1.89	1.27
4	2.42	2.16	2.29	.81	1.40	3.29	1.44
5	3.43	2.99	3.21	.92	1.84	5.14	1.60
6	4.53	3.87	4.20	.99	2.27	7.40	1.76
7	5.64	4.76	5.20	1.00	2.65	10.05	1.93
8	6.74	5.61	6.18	.98	3.02	13.07	2.12
9	7.89	6.42	7.11	.93	3.34	16.41	2.31
10	8.80	7.16	7.98	.87	3.63	20.04	2.51

*Feed Conversion = Pounds of feed consumed divided by the live weight of the birds.

of broiler production are shown in Table 16.

Most meat-type birds are sold as day-old chicks. A possible exception is the started capon, which may be purchased at four to five weeks of age from a hatchery or dealer.

TABLE 15
WATER CONSUMPTION OF BROILERS

GALLONS WATER PER 100 BIRDS

AGE IN WEEKS	WEEKLY	DAILY
0–1	7.0	1.0
1–2	10.3	1.5
2–3	17.3	2.5
3–4	22.5	3.2
4–5	25.7	3.7
5–6	30.2	4.3
6–7	34.8	5.0
7–8	38.8	5.5

Source: 1972-73 *New England Poultry Management and Business Management Manual*, Bulletin 566 (Revised).

TABLE 16
ESTIMATED PER-BIRD COSTS OF BROILER PRODUCTION

ITEM	COST
Chick Cost	$ .35-.40
Brooding	.10
Feed (7.4 lbs. at .10)	.74
Litter and Misc.	.10
TOTAL PER BIRD	$1.29-1.34

HOUSING AND EQUIPMENT NEEDS

Meat-type birds can be grown either in confinement or on range. Most broilers are grown in confinement. For at least the first several weeks they have to be confined to the brooder house, and since the males can be dressed as early as the sixth week, it usually doesn't pay to move them to range. The best growth results are obtained by rearing in confinement.

Roasters and capons are sometimes grown on range or yarded. This is satisfactory where housing is scarce, but growth and finish may be sacrificed by doing this. Meat birds usually do better when confined and receiving a full feed of a specialized ration designed for rapid growth and efficient conversion.

FLOOR SPACE

Meat birds should receive at least ½ square foot of floor space per bird until they are two weeks old and 1 square foot per bird between the second and tenth weeks. Heavy meat birds such as capons and roasters from ten to twenty weeks of age should have 2 to 3 square feet of floor space per bird. When meat birds are kept over twenty weeks they should have a minimum of 3 square feet per bird and more if available.

FEED AND WATER SPACE

The chicks may be fed in box lids for the first five to ten days, or chick-sized feeders may be used at the rate of 1 inch per chick. At three to six weeks, they should have 2 inches of feeder space per chick and at seven to twelve weeks, 3 inches. Birds to be carried on as roasters should be provided with 4 inches of feeder space per bird for the remainder of the growing period.

Provide water at the rate of 20 inches of hopper space or two 1-gallon waterers per one hundred chicks up to three weeks of age. From the third week they should have a 4- to 5-foot trough per one hundred chicks, or two large fountain-type waterers.

The feed hopper should not be filled more than half full to avoid feed waste. Adjust feeders so that the lip of the hopper is at a level with the bird's back. Adjustable hanging tube feeders are excellent for growing birds. Three tube-type feeders will normally accommodate one hundred birds.

HOUSING REQUIREMENTS

General housing requirements—the requirements for cleanliness, sanitation, litter, the arrangement of feeding and watering, and brooding equipment—are all much the same as those described for brooding replacement pullets in Chapter 4. Roosts are not recommended for brooding meat birds. Roosting can cause such problems as crooked breast bones and breast blisters.

Lighting meat birds is somewhat different from lighting replacement birds. Enough light should be provided the broiler chick to enable it to move about and see to eat and drink. Actually, it is desirable to keep activity at a reduced level for the most efficient feed utilization. The intensity of illumination at the level of the chick should be about ½ footcandle. It is sometimes difficult to keep this level of lighting in housing with windows. Birds that receive a higher intensity are more prone to feather picking and cannibalism, increased activity, and possible piling and smothering. One bulb watt per 8 square feet of floor space will normally provide ½ footcandle intensity. This is when the bulb has a reflector and is 7 to 8 feet above the floor.

Many broiler producers use a continuous-light program. This may be somewhat hazardous. In the event of a power failure, the birds may panic and pile. A light day of fourteen hours is usually sufficient. If the birds are given enough light to enable them to consume adequate feed and water, it is not usually necessary to provide more than fourteen hours. When the weather is exceptionally hot, it may be an advantage to use a sixteen-hour light day to provide additional feeding time during the cooler hours of the morning or evening.

MANAGEMENT SYSTEMS FOR MEAT BIRDS

Although some research work is being done to develop cage equipment for broilers and other meat birds, the recommended management system is still the conventional floor-litter system. Cage growing offers several advantages but, thus far, the disadvantages outweigh these advantages. More breast blisters, leg and foot problems, crooked keels, poor growth, and feed efficiency are problems that may result with cage-management systems of growing meat birds.

The windowed house or controlled-environment houses are used for meat birds depending upon the climate, the size of flock, and other factors. Ventilation and ventilation systems are the same as those for other types of growing houses. In most areas concrete floors are required for proper cleaning and sanitation and to avoid disease and parasite problems. The cleaning and sanitation requirements are the same for meat birds as for other growing stock.

DISEASE CONTROL

Disease prevention is vitally important to a broiler growing program. Unlike birds grown for egg production, or meat birds such as the roasters or capons, broilers have only an eight-week growing period. If the birds become sick in the middle of the growing period, the time in which the disease can be treated and brought under control is relatively limited, and the overall flock results can be severely affected. It is, therefore, important that a disease control program for broilers be one of prevention rather than treatment.

Vaccination recommendations for meat birds vary with the area. It is probably good to have the chicks vaccinated for Marek's disease, especially if the disease has been prevalent in your area. Marek's vaccine is usually administered at the hatchery. Some broiler growers do not vaccinate for other diseases because they practice isolation and sanitation practices and feel that these programs will prevent most disease outbreaks.

Before establishing a vaccination program for meat birds, be sure to check with poultry pathologists or other people who are knowledgeable about the possible disease exposure situation in the area. When broilers or other meat birds are vaccinated for something other than Marek's disease, it is normally Newcastle disease or infectious bronchitis.

It is very important to prevent outbreaks of coccidiosis in meat birds, particularly during the broiler stage. Outbreaks of coccidiosis normally occur early in the chick's life and can damage the intestinal lining. This prevents the normal absorption of food materials in the intestinal tract. Outbreaks of coccidiosis can cause poor growth and feed conversion, or—in severe cases—mortality and morbidity.

Drugs called coccidiostats are normally used in the feeds to either reduce the infection of coccidiosis or to completely suppress it. When pullets are grown for a floor management system, the starter feeds and grower feeds contain low levels of a coccidiostat, which permits the birds to have a low level of infection and build up immunity to the disease. Birds that are grown on the floor and are to be housed in cages—where they won't be exposed to fecal material and contained sporulated oocysts—are given a type of coccidiostat, or dosage of coccidiostat, in the diet to prevent infection altogether.

Since the growing period of broilers is so short, and so much emphasis is placed on fast, efficient growth, the usual procedure is to control coccidiosis during the growing period. Better growth results are obtained if infection is prevented. Most roaster and capon growers use a coccidiostat throughout the growing period.

Even if the birds are on a preventive level of coccidiostat, at times there are outbreaks of the disease due to excessively wet litter or other factors. The method of treating these outbreaks is covered in the section on flock health, Chapter 11.

FEEDING PROGRAMS FOR MEAT BIRDS

The goal of the feeding program for meat birds is to get as much growth as quickly as possible. This is unlike feeding replacement pullets where we are interested in delaying sexual maturity. Meat birds must have access to full feed at all times, and they should be encouraged to eat as much as possible, with as little waste as possible. A number of small-flock owners permit capons and roasters to yard or range. Though overall performance may be somewhat less, excellent birds can be grown this way. The cost may be somewhat less also, because the chickens forage for a part of their food intake.

It is important to feed meat birds a ration designed to give rapid growth and good feed conversion. They should be started on a broiler feeding program. The complete feeding program includes two or possibly three different rations. Some companies recommend a program that begins with a broiler starter the first five weeks. The starter contains about 22 percent protein and a coccidiostat. This feed is replaced with an 18 percent broiler finisher from five weeks up to five days prior to marketing. At this time they would be shifted to a final feed of 17 percent, containing no coccidiostat. Where this feeding program is used, the birds must be checked very carefully for coccidiosis if they are to be held beyond the five-day period. It may become necessary to treat for coccidiosis if the birds are held more than five days prior to dressing or marketing.

If birds are to be held over as light roasters for four to five weeks beyond the broiler period, they can be fed a 20 percent broiler finisher diet. This may be fed up to two to five days prior to dressing or marketing, depending upon the type of coccidiostat used. Some coccidiostats must be withdrawn several days prior to slaughter to avoid carcass residues.

If birds are fed as capons or heavy roasters, they should be fed a lower-protein and higher-energy diet from the ninth or tenth week to marketing. The protein level should be approximately 16 percent. Some feed company programs call for feeding a roaster finished from the sixth week to two days before marketing, at which time a feed containing no coccidiostat is used for the remainder of the period.

CAPON PRODUCTION

Male chickens castrated or caponized to make them fatten more readily are called capons. Capons are not only more tender than cockerels but bring a much better market price. Capons are normally grown to be marketed during the Thanksgiving and Christmas holiday seasons, at which time they bring premium prices.

Up to possibly twenty weeks of age the capon and the cockerel weigh about the same. Afterwards, the capons gain weight more rapidly than cockerels. When cockerels show spur and comb development they are called stags, and their flesh toughens. Toughness of the flesh does not occur in capons. When cockerels are a year old, and are classed as old cocks, they bring a very low price and their flesh is very tough. On the other hand, capons have very tender flesh, and if kept beyond the age when they are normally marketed or dressed off, the flesh remains tender.

Unlike the cockerel, the capon has a very quiet disposition, is very docile, and seldom crows. After caponization the comb and wattles cease growing, the head looks pale and small, and the hackle, tail, and saddle feathers grow to be unusually long.

SELECTING A BREED FOR CAPONIZING

Broiler stocks with white plumage and yellow skin are preferred for meat bird production. White Plymouth Rocks, New Hampshires, and crosses of these breeds, or those crosses involving White Cornish and White Rock Crosses, are generally used. Caponizing small birds, such as Leghorns or the small representatives of the American breeds, just does not pay.

WHEN TO CAPONIZE

Since capons are in the biggest demand during the Thanksgiving-to-March period and take twenty-four to twenty-five weeks to grow and finish properly, the best time to caponize is in late spring or early fall.

Caponizing used to be done at five to six weeks of age. With our faster-growing broiler strains of birds, it is now desirable to caponize at approximately twelve to twenty days of age before they become too large, when the operation is more difficult. Some hatcheries or dealers specialize in selling four-week-old capons that have been caponized as early as ten days. It is possible to caponize older birds up to eight weeks; it is best, however, to do it earlier.

THE CAPONIZING OPERATION

Feed and water should be withheld from the birds for twenty-four hours before the operation. The bird's intestines should be almost empty to prevent them from obstructing the operation. When the birds are removed from feed and water, keep them in wire or slat-bottom coops to prevent them from eating litter or other materials.

Good light is mandatory during the operation. Direct sunlight is best; a strong electric light with a reflector can, however, serve indoors.

A barrel or box can be used as an operating table. When a large number of birds are to caponized, a table of convenient height is recommended.

During the operation, the birds can be restrained by a second operator who holds each bird outstretched on its side. A caponizer can do the operation single-handedly by restraining the birds with weighted cords tied to the legs and wings. These cords should be about 2 feet long and weighted with 1-pound weights. The cockerel's legs are securely tied by the use of a half hitch in one cord. Both wings are held together near the shoulder joint with a half hitch in the other cord. The weights are hung over the edge of the operating table, and the bird is stretched out. The cords should be arranged so that it is easy to adjust the bird and turn it over without disturbing the weights.

METHODS

The operation can be done with one incision or two incisions. Most operators find it easier to remove the upper, or nearer testicle, then turn the bird over and make a second incision on the other side of the body to remove the other testicle. If both testicles are removed through just one incision, it is best to remove the lower one first; otherwise, bleeding from the upper may obscure the lower.

It is very important not to rupture the artery that runs just behind the testicles, or the chicken will bleed to death in seconds. If feed and water have been withheld, the artery usually presents no problem.

The instruments needed for the operation are: a sharp knife, a rib spreader, a sharp pointed hook, and the testicle removers or forceps (Figure 23). The operation should be performed as quickly as possible to reduce stress on the bird.

Moisten and remove the feathers from a small area over the last two ribs, just in front of the thigh; with one hand, slide the skin flesh down toward the thigh and hold it in that position. This will prevent cutting into the muscle and causing

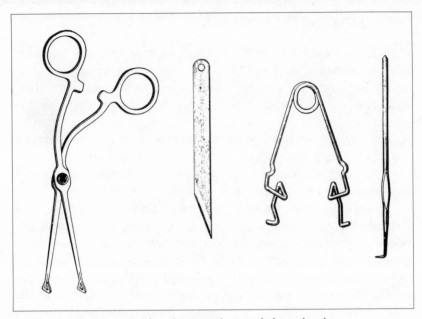

Figure 23. Forceps, knife, rib spreader, and sharp hook.

excessive bleeding. The cut is made between the last two ribs (Figure 24). The incision should be lengthened in each direction until it is about ¾ to 1 inch long.

Then insert the spreader into the incision to spring the ribs apart (Figure 25). The intestines will be visible beneath a thin membrane. Tear open this membrane

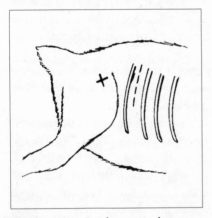

Figure 24. Cut between last two ribs.

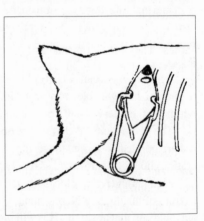

Figure 25. Testicle exposed.

with the hook to expose the upper testicle. It is usually yellow in color but may be dark colored. It is about the size and shape of a navy bean and is located up close to the backbone and just below the front end of the kidney. If the intestines are empty, once they are pushed aside it is quite easy to see the lower testicle. It is in the same position as the upper one, but on the other side of the backbone.

The forceps are used to grasp the testicle and it is important that the artery be avoided at this point. The entire testicle must be removed with a slow twisting motion, tearing it away from the spermatic cord to which it is attached. Unless both testicles are removed with the same incision, the bird is turned over, and the same operation is performed on the other side. It is not necessary to stitch the incision, since when the skin and flesh are released they will slide back over the incision and cover it. The caponizing tools should be rinsed, after each bird is caponized, in a solution consisting of ten drops of Clorox or chlorine bleach per quart of water.

SLIPS

Frequently, when the caponizing operation is performed for the first time, a number of birds will become what are known as *slips*. A slip is neither a cockerel nor a capon; it develops when a part of the testicle is not removed during the operation. This small piece of testicle often grows to a considerable size.

Slips have the same restless disposition as cockerels and they grow and fatten little, or no better, than a cockerel. Thus slips are not as highly valued for meat purposes and do not bring as good a price if sold. Inexperienced operators may expect up to 50 percent slips; experienced operators can expect about 5 percent or less.

Slips develop red heads and wattles whereas a true capon has a pale head and wattles. They can be finished early and marketed as broilers at around eight to nine weeks of age. They may be kept to twelve to thirteen weeks of age and sold as roasters. Older birds develop tougher meat.

LOSSES IN CAPONIZING

Occasionally, even the best operators will kill some birds during the caponizing operation. However, this loss seldom exceeds 5 percent. Inexperienced caponizers frequently kill several birds, but gradually losses are reduced as they gain experience. Any birds killed during the operation can be dressed and eaten.

Speed develops with experience. An experienced caponizer can do up to four birds per minute. It is wise for the beginner to practice on slaughtered birds to acquire some skill before attempting to caponize a live bird. Caponizing is somewhat of a stress on the birds, so some commercial caponizers inject the birds with an antibiotic solution just prior to the operation. Some capon growers administer the antibiotic solution in the drinking water for two or three days, beginning four or five days prior to the operation. Antibiotics should not be given by way of the drinking water within two days of the operation. Some antibiotics tend to cause the intestines to balloon, thus increasing the possibility of damage to the intestine and making it more difficult to spot the testicle.

CARE AND FEEDING OF CAPONIZED BIRDS

Caponized birds should be separated from other chickens and kept separated during the growing period. Air puffs or wind puffs may develop after the operation. Wind puffs are caused by air that gathers under the skin. This air should be released by pricking the skin with a needle or knife and pressing it out. Usually within ten days after the operation the incision is fully healed.

Breast blisters are often a problem with capons. They are likely to appear when the capons are about half grown, and the incidence may increase as the birds become heavier. Birds may develop breast blisters from roosts, especially those with narrow perches, or from resting on board, concrete, or wire floors. Litter should not be permitted to become caked or excessively wet but should be kept soft, loose, and dry. Roasters or capons should not be permitted to roost.

During the growing period, the birds should be supplied with a growing feed containing approximately a 17 percent protein. The growing stage is very important because during this period leg problems may occur. Capons, if pushed too hard for rapid growth during the early growing period, may have insufficient bone development to support their body weight. Most cockerels that are to be used for capons are actually broiler males. This stock has been selected for rapid growth and males should reach 3½ to 4 pounds in body weight at eight weeks of age. They are normally marketed at these weights. When we use these birds for capon production it is expected that they will grow to about 9 or 10 pounds live weight in approximately twenty weeks. Some of the birds will not have strong enough legs to support these heavier weights if growth is too rapid early in the growing period. It is, therefore, well to keep this in mind when selecting a feeding program for capons. Some growers restrict feed, and thus the growth rate, from about six to ten weeks of age to prevent this problem in light roasters. Feeds specially formulated for capons take this problem into consideration.

TABLE 17
APPROXIMATE COSTS OF PRODUCING CAPONS
AND ROASTERS

Item	Cost
Chicks	$ .35–.85
Brooding, litter, medication, misc.	.20
Feed (24–38 pounds) @ $.12	2.88–4.56
Surgical caponizings	.00–.25
TOTAL	$3.43–5.86

Feed cost will vary with the age at which birds are slaughtered. Those slaughtered as light roasters (twelve to thirteen weeks of age) will consume less feed than capons. Roasters are not normally caponized, thus no caponizing charge. Chick costs will vary with source, size of order, whether sexed or straight run, and other factors.

Broiler rations, however, are high-energy feeds designed for maximum growth and may create the leg problems if fed throughout the growing period. For heavy roasters, it is suggested that a lower-energy diet be used from approximately eight to thirteen weeks.

Leg weaknesses also contribute to the breast blister problem. This problem can be very serious because it detracts from the dressed appearance of the birds. Birds that do have a weak-leg problem exhibit a tendency to sit down much of the time. The keel bone is constantly in contact with the litter and the incidence of blisters is increased, especially if litter conditions are poor. Other factors that affect the incidence of breast blisters include the type of equipment used, overcrowding, genetics, loading and handling methods, and others. Birds can receive bruises and skin irritations by bumping into equipment or by being pushed and shoved or mishandled.

Capons can be grown in confinement with a yard provided or on range. Downgrading often seems more common in the case of confinement-reared birds. The reason for this is that confined birds are usually pushed to higher weights more quickly. There is also a greater possibility of crowding and injuries when birds are confined.

Range birds, however, seem to have higher mortality. A lot of this is probably due to predator losses. Birds on range should be provided with shelters. Wire floors and perches are commonly used, but these may cause breast blisters. Wire floors and perches should be covered, and preferably litter used on the floor of the shelters. Shade is important for birds on range. This is especially true during the hot-weather seasons, or in warm-weather climates. The range should be clean and provide a short succulent forage to be most valuable. If good range is provided, feed costs can be substantially reduced. Birds on range should be provided with range feeders that prevent the feed from being blown away or getting wet and moldy. Adequate feeder and watering space is important regardless of whether the birds are reared on range or in confinement. Approximately 4 inches of feeder space and 1 inch of watering space per bird is required.

About two weeks prior to slaughter, capons should be put on a finishing diet to get them in good marketing condition in terms of fat covering, fleshing, and pigmentation. If pushed too early with finishing diets, the birds may become overly fat, go off their legs, and develop breast blisters. It doesn't pay to keep capons beyond twenty weeks for efficient feed utilization.

CHAPTER 8

Turkey Production

BEFORE YOU LAUNCH into the production of turkeys, even on a small scale, you should be aware of the costs of production. The turkey poult costs substantially more than does a chick and, of course, turkeys consume considerably more feed than do growing chicks. Thus the cost of producing turkeys is considerably higher. Table 18 presents the cost involved in raising a small flock of turkeys.

GETTING STARTED

The turkey enterprise can be started in one of several ways. You can purchase hatching eggs and incubate them or purchase day-old poults. One may also buy breeding stock, but this is a rather expensive way to begin. If hatching eggs are purchased, it requires equipment for incubating; if breeding stock is purchased, it is expensive not only to buy the birds but also to maintain them. Normally, most small flocks of turkeys are purchased as poults from a hatchery or agricultural supply store.

TABLE 18
ESTIMATED PER-BIRD COSTS OF RAISING
HEAVY ROASTER TURKEYS

ITEM	COST
Poult	$1.75–4.00
Feed (80 pounds) @ $.12–.15	9.60–12.00
Brooding, litter, medication, misc.	.25 .25
TOTAL	$11.60–16.25

Costs will vary with number of poults purchased and volume of feed purchased, as well as sources and other factors.

BREEDS

There are several varieties of turkeys, but those that are most important commercially are the Broad Breasted Bronze, the Broad Breasted White, and the Beltsville Small White. Until quite recently the Bronze was the most popular variety.

The Bronze has good growth rate, conformation, feed conversion, and about all of the qualities demanded by the turkey industry. Its basic plumage color is black, and it has dark-colored pinfeathers—a distinct disadvantage that has led to its replacement by the Broad Breasted White.

The Broad Breasted White was developed in the early 1950s from crosses of the Broad Breasted Bronze and the White Holland variety. In some areas of the country it is now difficult to find the Broad Breasted Bronze because the current demand is so great for the Whites.

The Small White looks very much like the Large White from the standpoint of color and body conformation. It is a much smaller turkey that was developed at the Beltsville Agricultural Research Center in Maryland. Its body weight has tended to increase in recent years through selection by breeders.

Properly managed and fed, the Beltsvilles reach good market condition in fifteen to sixteen weeks and make excellent turkey broilers. They also make good medium roasters if held to twenty-one to twenty-four weeks. They are not as efficient converters of feed as the Large Whites but usually cost somewhat less as poults. Their main disadvantage is the difference in feed required to reach the same weights as the large turkeys.

BUYING POULTS

When buying poults you should select a strain that is known to yield good results. As is the case with chickens, poults should originate from sources that are U.S. pullorum-typhoid clean, and preferably from breeder flocks having no history of sinusitis, or air-sac infection. Consult your state Poultry Specialist, your county Agricultural Extension Agent, or some other knowledgeable person for advice on sources of good turkey poults in your area.

Place your order for poults well in advance of the delivery date, so as to be sure to get the quality of stock you want. Birds should be ordered to arrive so as to permit twenty-four to twenty-eight weeks for growing the large breeds and eighteen to twenty-two weeks for smaller breeds. It is sometimes difficult to obtain small lots of turkeys delivered from the hatchery. However, it may be possible to pick the turkeys up at the hatchery or, perhaps, have a nearby producer order a few extra poults for you. Poults that are shipped express are subject to chilling and overheating. Buy the poults as close to home as possible.

HOUSING REQUIREMENTS

Small flocks of turkeys are normally started in the warm months of the year, so housing doesn't have to be fancy. The brooder house should be a well-constructed building that is easily ventilated, however. If a small building is not available, a pen can be built within a large building. The pen should have good floors that can be readily cleaned. The insulation required will depend upon the climatic conditions as well as the time of year the poults are to be brooded. A well-insulated building can save energy and brooding costs. Young turkeys must be kept warm and dry, so a well-insulated and ventilated house is important in cool climate, particularly with large flocks.

The brooder house should have windows that slide down or tilt in from the top. These are best for ventilation. Windows should be located in the front and back of the house. Usually 1 square foot of window area to each 10 square feet of floor area is adequate.

Confinement rearing of turkeys is most commonly used by the small grower. Where predators or adverse weather conditions are likely, or where there is limited range area, confinement rearing is preferred. Some producers yard their birds within fenced-in areas using the brooder house as the shelter.

Sun porches were once very popular with turkey growers, and some producers still use them. The porches are attached to the brooder house or shelter. The floor is made either of slats or wire. They are often elevated to provide a space underneath for the accumulation of droppings and easy access for cleaning. Porches are fenced in on the top and sides. Coarse mesh wire is used on the top to avoid snow loading in those areas where that may be a problem.

Porches have several advantages in that they help to acclimate the birds to the weather conditions prior to going on range. This helps to reduce stampeding on the range during wind or rain storms. Porches also acclimate them to changes in temperature to which they are exposed on range.

Heavy-breed turkeys today are not grown on wire or slated porches as much. Breast blisters, and foot and leg problems tend to develop particularly on wire. Light-breed turkeys can still be grown on porches. A modification of the porch approach is the use of paved yards or yards with a gravel or stone surface.

Large turkey production units sometimes use windowless, controlled-environment houses. These buildings are thoroughly insulated to provide

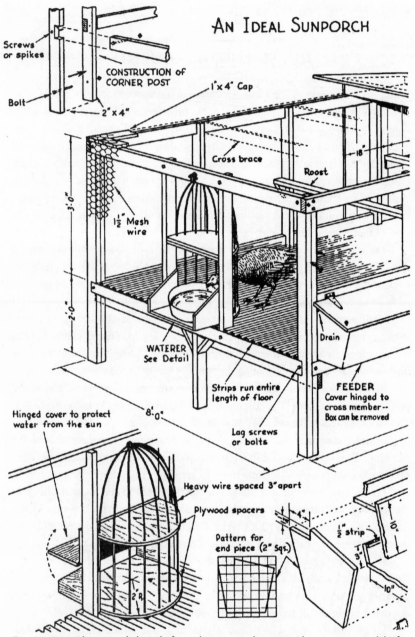

Figure 26. Plans and details for a house and sunporch system suitable for a small flock.

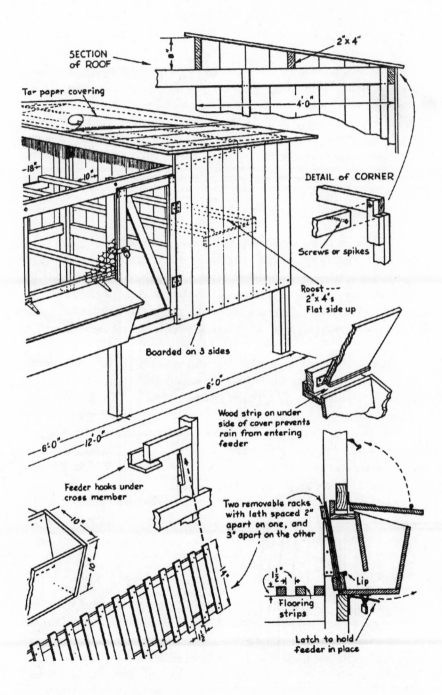

SECTION of ROOF

2" x 4"

4'-0"

Tar paper covering

DETAIL of CORNER

-18"

-10"

Screws or spikes

Roost ---
2" x 4's
Flat side up

Boarded on 3 sides

6'-0"

Wood strip on under side of cover prevents rain from entering feeder

6'-0"   -12'-0"-

Feeder hooks under cross member

10"

10"

Two removable racks with lath spaced 2" apart on one, and 3" apart on the other

Lip

1½"

Flooring strips

4"

1½

Latch to hold feeder in place

maximum heat efficiency during brooding and comfortable, well-ventilated conditions for the birds.

Ventilation systems for turkey housing are designed essentially the same as for chicken houses. Many of them use fan ventilation with the same type of inlet but a larger ventilating capacity. Fan systems are designed to move from ¾ to 1 cubic foot per minute at ⅛ inch static pressure per pound of turkey expected at maturity. If you plan to go into the turkey business on a commercial scale, contact your state Extension Poultry Specialist for help in designing the facilities.

PREPARING THE BROODER HOUSE

The brooder house should be cleaned between broods. This means completely removing all the old litter and thoroughly washing the floors, sidewalls, ceilings, and equipment. The building and equipment should be disinfected with a material such as cresilic acid. Some of these materials can cause foot burn or eye injury. To prevent disinfectant injury to the poults, make sure that the house has an opportunity to dry for about two weeks prior to the time the poults are put in the house.

After the building is thoroughly dry, put 3 to 4 inches of litter down on the floor. Some of the common litter materials suitable for use are wood shavings, sugarcane, ground corn cobs, peatmoss, or vermiculite. The litter may be covered or uncovered. Some producers cover the litter with paper to prevent litter eating for the first week. If the litter is covered, use a rough paper, as a slippery surface can cause leg weakness and crooked feet. The litter should be evenly distributed over the floor and free of mold and dust. Very coarse litter material can also contribute to leg disorders.

In small brooder houses, corners should be rounded with small mesh wire to prevent piling. Turkeys may pile when frightened or when the house is drafty or floor temperatures low.

FLOOR SPACE

Use 1 square foot of floor space per poult up to eight weeks of age. From eight to twelve weeks the floor space should be increased to 2 square feet per poult, and from twelve to sixteen weeks 2½ square feet should be the minimum allowance. Mixed sexes kept in confinement during the entire growing period should receive 4 square feet per bird. If the flock is all toms, provide 5 square feet or, if all hens, 3 square feet. For light turkeys or for turkeys housed in controlled-environment housing, floor space requirements may be somewhat less than the above recommendations.

BROODERS

Several types of brooders are suitable for brooding poults. Gas or electric are probably the best suited for small-flock situations. If hover-type brooders are used, allow approximately 12 to 13 square inches of hover space per poult started. A brooder that is rated for 250 chicks will be adequate for only about 125 turkeys. When hover-type brooding is used, a 7 ½-watt light bulb should be used to attract the poults to the heat source. Each brooder should be equipped with a thermometer that is easily read. Infrared brooders are satisfactory for small turkey flocks.

If infrared bulbs are used for brooding, provide two or three 250-watt bulbs per 100 poults. Infrared lamps should be hung 18 inches from the surface of the litter initially. They can be raised 2 inches per week until they're 24 inches above the litter. The brooder ring or guard should be 8 to 10 feet in diameter. Take care to avoid splashing water on infrared bulbs. This may cause them to break unless they are the Pyrex type. Ideally, the temperature outside the heat source or the hover area should be approximately 70° F. to provide maximum comfort for the poults.

BROODER GUARDS

Use brooder guards to confine the birds to the heat source and to prevent drafts on the poults.

Management of the brooder guard varies, depending upon the design of the house and the climatic or seasonal conditions. Where noninsulated housing is used during fairly cool weather, an 18-inch brooder guard for each stove is recommended. For warm-weather brooding, a 12-inch brooder guard is satisfactory. The brooder guard should be located 2 to 3 feet from the edge of the hover and be gradually moved out to a distance of 3 to 4 feet. After seven to ten days the brooder guard can be removed. At this time the poults are normally allowed free access to the brooder house. If poults are brooded early in the season or in cold climates, and the brooder house is quite large, it may be necessary to partition off part of the house at the start and enlarge the area as necessary.

FEEDING EQUIPMENT

The first feed for the poults can be supplied to them on new egg filler flats, on chick box lids, or in small chick feeders. Turkeys sometimes have a visual problem and have difficulty finding the feed and water. Starve-outs or dehydrated poults may result. Bright colored marbles placed on top of the feed or in the water containers will often attract the poults to the feed and water. Oatmeal or fine granite grit sprinkled very lightly over the feed, once or twice a day for the first three days, may also help to get them eating.

Unless the litter is covered with a paper, don't fill the feeder so full that it will overflow. This can lead to the practice of eating litter. For poults from seven days to three weeks old, use small feeders at the rate of 2 inches of feeder space per bird. From three weeks to market, the poults should be furnished with large feeders about 4 inches deep at the rate of 3 inches of feeder space per bird. Hanging tube-type feeders can also be utilized. The amount of tube-type feeding space can be determined by multiplying the diameter of the feeder pan by 3-1/7 to obtain the number of inches of feeder space available. In figuring feeder space remember to multiply the hopper length by two if the poults are able to use both sides of the feeder. Thus a 4-foot trough feeder actually provides 8 linear feet of feeder space.

WATER SPACE

Poults can be started on either glass fountain-type waterers or automatic waterers. From one day to three weeks the poults should have access to 1- or 2-gallon fountains per one hundred poults. From three weeks to market they should have 3- or 5-gallon fountains per one hundred poults, or one 4-foot automatic waterer.

Changes of equipment, both feeders and waterers, should be made gradually so as not to discourage feed or water consumption.

BROODING THE POULTS

When the poults arrive the brooder house should be completely ready for them. The feeding equipment should be arranged and the brooder guards set up (see Figure 10, Chapter 4). The brooders should have been in operation for approximately twenty-four hours to get them regulated and to get the building warm.

If hover-type brooders are used, the temperature should be 100° F. for White birds, 95° F. for Bronze. This temperature reading is taken under the hover approximately 7 inches from the outer edge and 2 inches above the litter, or at the height of the poults' backs. Be sure to check the accuracy of the thermometer you are using before the poults arrive. The hover temperature should be reduced 5° weekly until it registers 70° or 75° F., or equals the outside temperature. If the weather is warm during the brooding period, heat may be shut down during the day after the first week. Heat during the evening hours will be required for a longer period. Normally, little or no heat is required after the sixth week, depending upon the time of year, the weather conditions, and the housing. Watch the poults as a guide when checking or adjusting temperatures (Figure 11, Chapter 4).

Important: Feed and water the poults as soon as possible after hatching. As each poult is taken from the box, dip its beak in water and then in the feed. This will help to get it started on feed and water.

LIGHTING

High light intensity should be provided the poults for the first two weeks of brooding in all types of houses to prevent starve-outs. Ten to 15 footcandles of light should be used day and night. This will require 200-watt bulbs spaced 10 feet on centers. A small 7½- to 15-watt attraction light bulb should be installed under each brooder hover. After the first two weeks in window houses, use dim lights at approximately ½ footcandle during the night hours. The dim lights will help to prevent or discourage piling and stampeding.

ROOSTS

Roosts are seldom used for brooding turkeys though they may help to prevent piling at night. Roosts do not normally cause breast blister problems with turkeys. If roosts are used in the brooder house they may be of the step ladder type, allowing about 3 linear inches of perch space per bird at the start. The perches are made of 2-inch-round poles of 2 x 2 or 2 x 3 material, with the lowest perch located about 12 inches above the litter. Each succeeding perch is 4 to 6 inches higher. Perches should be beveled or rounded on the edges to prevent injury to the breasts. Each bird should have 6 linear inches of roosting space by the end of the brooding period. The bottom and the ends of the roosts should be screened so as to prevent the poults from gaining access to the droppings. Birds usually begin to use the roosts at about four to five weeks of age.

If birds are grown in complete confinement, roosts usually are not used. Since the birds bed down on the floor, litter conditions should be kept in good condition to prevent breast blisters, soiled and matted feathers, and off-colored skin blemishes on the breast. Good litter conditions also help to maintain sanitary conditions and prevent disease problems.

Roosts used on range should be constructed of 2-inch poles or 2 x 4 material laid flat with rounded edges. Perches should be spaced 24 inches apart and located approximately 3 feet off the ground. If placed in a house or shelter, roosts can be slanted to conserve space. For outdoor roosting racks, all roost perches should be built on the same level. This type of roost should be built of fairly heavy material to prevent breaking when the weight of the birds is concentrated in a small area. Ten to 15 inches of perch space per bird is required for the large birds, and 10 to 12 inches for small birds.

FEEDING

One of the soundest pieces of advice that can be given to a turkey grower is to select a good brand of feed and follow the manufacturers' recommendations for feeding.

Two basic feeding programs for growing turkeys are available. One is the all-mash system and the second is a protein-supplement-plus-grain system. Nutrient requirements of turkey poults vary with age. As birds become older, protein, vitamin, and mineral requirements decrease and energy requirements increase.

Feed an insoluble grit such as granite grit the first eight to ten weeks. Whenever grains are included in the diet, or when the birds are on range, feed insoluble grit to enable the birds to grind and utilize the grains and fibrous materials.

Recommendations for feeding, and the number of feeds offered, vary considerably among feed companies. One of the simpler types of feeding programs recommends a 28 percent starting diet with a change to a 21 percent growing diet, and finishing with a 16 percent protein diet. Other commercial concerns offer and recommend five or six different diets during the growing period. Again, buy feed from a reputable feed company and follow their recommendations. Under most circumstances the diets should include a coccidiostat and, in most instances, a blackhead disease preventative drug should be incorporated in the feed.

Feed and water should be kept before the birds continuously. Pelleted mash can be fed after the first four weeks. Most growers feed a nutritionally complete starting mash; however, green feed can be utilized for small flocks where labor requirements are of little concern. Tender alfalfa, white Dutch clover, young tender grass, or green grain sprouts all chopped into short lengths and fed once or twice daily are good for the poults. Wilted or dry roughage feeds should not be permitted to remain before the poults. These can cause impacted or pendulous crops.

Turkey starter diets can be purchased ready-mixed. They can be home- or custom-mixed according to recommended formulas, or a concentrate can be used and mixed with ingredients such as ground corn and soybean meal. For the most part, small producers will find it best to feed a ready-mixed complete starter diet.

Growing diets, fed to the poults from eight weeks to maturity, come in mash or pellet form. They may include either loose or pelleted mash plus whole or cracked grain, sometimes referred to as scratch grains. A commercial concentrate may also be combined with ground grain or with soybean meal and ground corn in the proportions recommended by the manufacturer. Once again the small grower will find it advantageous to use a complete ready-mixed mash when the birds are reared in confinement. If the birds are on range, a complete feed, preferably in pellet form, supplemented with good range, grains, and insoluble grit, makes a sound feeding program. Growth rate and feed consumption information for heavy and light roaster turkeys is presented in Table 19. Water consumption information is presented in Table 20.

RANGE REARING

It is possible to reduce the cost of rearing turkeys by putting them on range. This is especially true if the diet can be supplemented with home-grown grains. Turkeys are good foragers, and if good green feed is available on the range this will help to supplement the diet and reduce the cost of the feeding program.

Range rearing is not without its problems. Losses are possible from soil-borne diseases, insects, thieves, adverse weather conditions, and predators. Because of these factors confinement rearing, which tends to minimize these losses, has quite rapidly replaced range rearing in some areas.

Portable range shelters that can be moved to various areas of the range are preferred with layer flocks. The shelter should provide about 1½ square feet of

TABLE 19
GROWTH RATE AND FEED CONSUMPTION OF
RAPID-GROWING ROASTER TURKEYS

AGE (WEEKS)	LIVE WEIGHT (POUNDS)		TOTAL CUMULATIVE FEED REQUIRED (POUNDS)	
	Toms	Hens	Toms	Hens
2	.64	.60	.75	.70
4	1.80	1.67	2.56	2.29
6	3.80	3.35	5.80	5.26
8	6.50	5.55	11.43	10.05
10	9.76	8.05	19.30	16.54
12	13.48	10.77	29.66	24.74
14	17.59	13.39	42.77	34.00
16	21.84	15.86	58.15	43.77
18	26.07	18.07	75.43	54.93
20	30.25	20.00	93.87	67.80
22	34.24		115.19	
24	37.97		137.61	
TOTAL	**37.97**	**20.00**	**137.61**	**67.80**

Based on material by Jerry L. Sell, Department of Animal Science, Iowa State University.

floor space per bird. It should be equipped with roosts, or slat floors with the slats located 1½ inches apart.

Normally, turkey poults can be put out on the range at eight weeks of age. The flock should be well feathered, especially over the hips and back, before they are put on range. Check the weather forecast for several days before putting the birds on range. It is best to move them in the morning, as this helps to get the birds used to range conditions without losses.

A range area that has been free of turkeys for at least one year, and preferably for two years, is desirable. Poorly drained soil on the range should be avoided, since stagnant surface water may be a factor in turkey disease outbreaks. A temporary fence can be used to confine the flock to a small area of the range. Move the fence once a week, or as often as the range and weather conditions indicate. Provide artificial shade if there is no natural shade. Several rows of corn planted along the sunny side of the range area provides good shade and will also

TABLE 20
DAILY WATER CONSUMPTION OF HEAVY
ROASTER TURKEYS
(GALLONS PER 100 BIRDS)

Age in Weeks	Water Consumption	Age in Weeks	Water Consumption
1	1	11	14.0
2	2	12	15.0
3	3	13	16.0
4	4	14	16.5*
5	5	15	17.0*
6	6	16	16.5*
7	7.5	17	16.5*
8	9.5	18	16.5*
9	11.0	19	16.5*
10	12.5	20	16.5*

Source: Dr. Salsbury Laboratories

*Consumption will vary from 15 weeks to maturity from 14 to 19 gallons per 100 birds per day depending upon environmental temperatures.

provide some feed as it matures. About 1 acre of good pasture is required for 250 turkeys. If range shelters are used, it is best to move them every seven to fourteen days depending upon the weather and the quality of range. Feed and watering equipment should also be moved as needed to avoid muddy and bare spots.

It is possible to use a combination of confinement and range rearing by providing a permanent shelter such as a barn with fenced-in range areas around the barn. These fenced-in areas can be used on an alternative basis every year or two.

Range crops will depend upon the climate, soil, and range management. Many turkey ranges are permanently seeded; others are a part of a crop rotation plan. As part of a three- or four-year crop rotation, legume or grass pasture and an annual range crop such as soybeans, rape, kale, sunflowers, reed canary grass, and sudan grass, have been used successfully. Sunflowers, reed canary grass,

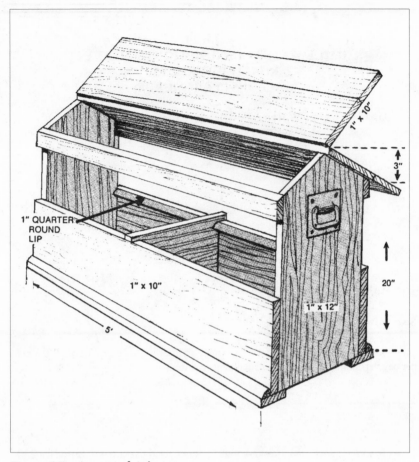

Figure 27. A range feeder.

and sudan grass provide green feed and shade. For permanent range, alfalfa, ladino clover, bluegrass, brome grass, and others are very satisfactory.

If feeders are to be used outside, they should be waterproof and windproof so that the feed is not spoiled or blown away (see Figure 27). The feeders should be placed on skids, or be small enough to move by hand. Trough-type feeders are inexpensive and relatively easy to construct. Specialized turkey feeding equipment can also be purchased. To prevent excessive feed waste, all feeding equipment should be designed so that it can be adjusted as the birds grow. The lip of the feed hopper should be approximately on line with the bird's back and the feed level in the hopper kept at about half to prevent waste. Pelleted feeds

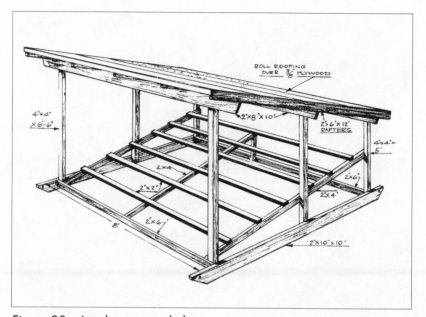

Figure 28. A turkey range shelter.

are less likely to be wasted on range.

Provide two 10-foot feeder troughs per one hundred birds if the flock is hand fed each day. When bulk feeders are used, feeder space should conform to the equipment manufacturer's recommendations. Provide one 4-foot automatic trough waterer or the equivalent per one hundred birds. The waterers should be cleaned daily and disinfected weekly, and, if possible, they should be shaded with portable or natural shade.

Range shelters should provide roosting space and cover for the birds during adverse weather and hot sun (Figure 28). During the hot period of the year, particularly when birds are close to marketing time, it is advisable to pull extra shelters onto the range to provide more shade for the birds when natural shade is not available.

GROOMING

The following grooming procedures are often performed to prevent injury to birds and thus improve their condition when they go to market.

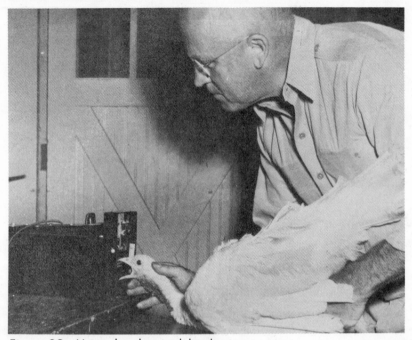

Figure 29. Using the electric debeaker.

DEBEAKING

It is recommended that the poults be debeaked when they are from two to five weeks of age. Delaying beyond this period makes it difficult to handle the heavier birds and excessive feather picking may occur. To prevent cannibalism and feather pulling, debeaking should be regular practice before noticeable picking occurs.

The poults should not be debeaked at day of age at the hatchery, as this can interfere with their ability to eat and drink and may cause starve-outs and dehydrated birds. The debeaking job is done with an electric debeaker and about half of the upper beak should be removed, being careful not to cut into the nostril. Make sure that feed and water levels are deep enough for debeaked birds. See Figures 30 and 30a for the correct debeaked appearance of turkeys.

Figure 30. Properly debeaked young turkey.

Figure 30a. Properly debeaked adult turkey.

DESNOODING

The snood (a fleshy appendage on the top of the head/back of the beak) is sometimes removed. Removal of the snood helps to prevent head injuries as a result of fighting or picking, and may prevent erysipelas infection from getting started in the flock. The snood can be removed at day old by pinching between the finger and thumbnail. It may be cut with fingernail clippers or scissors up to approximately three weeks of age.

WING CLIPPING OR NOTCHING

The wing feathers of one wing are sometimes cut off with a sharp knife to prevent the birds from flying, if on range. Wing notching, or the removal of the end segment of one wing with a debeaker, is another method sometimes used to prevent flying on range. Wing notching can be done with an electric debeaker from day of age up to ten days of age. Wing clipping and wing notching are not used as much as they once were because of injuries resulting when the birds attempt to fly. This may cause carcass bruises and detract from the dressed appearance.

TOE CLIPPING

Toe clipping is frequently done to prevent scratches and tears of the skin on the birds' backs and hips. It is especially helpful where birds are crowded or nervous, but appears to help even when birds are on the range. The two inside toes are clipped so that the nails are completely removed. Surgical scissors or an electric debeaker may be used. This should be done at the hatchery.

GENERAL MANAGEMENT RECOMMENDATIONS

Young poults should be isolated from older turkeys. Take care not to track disease organisms from the older stock to the younger stock. Follow a good control program for mice and rats. These rodents are not only disease carriers but also consume large quantities of feed. Rats can kill young poults.

Ideally, no other avian species such as chickens, game birds, or waterfowl should be on the same farm. However, with the medications available today, it is possible to grow small flocks of turkeys where other types of birds are kept, if they are kept separate. It is still somewhat risky. If abnormal losses or disease symptoms occur, birds should be taken promptly to a disease diagnostic laboratory for diagnosis. One of the first symptoms of a disease problem is a reduction in feed and water consumption. Without good records on daily feed consumption it is difficult to detect changes in feed intake. Dead birds should be disposed of immediately in a sanitary manner, by incinerating or by use of a disposal pit (see Chapter 11).

With good management, the turkey grower should be able to raise 95 percent of the turkeys started to maturity. With high poult costs and feed costs, mortality can become very costly, especially when birds are lost towards the latter part of the growing period.

Depending upon the disease exposure in a given area it may be necessary to vaccinate the poults for such diseases as Newcastle, fowlpox, erysipelas, or fowl cholera. In planning a vaccination program for your flock, check with the poultry diagnosticians at your state laboratory, your county Extension Agent, or some other knowledgeable person. Small flocks often are not vaccinated.

Medication is effective in reducing losses from such diseases as coccidiosis and blackhead when used in the feed at preventive levels. Antibiotics and other drugs are of value in preventing and treating diseases, but such medications should not be used as a substitute for good management.

THE PRODUCTION OF TURKEY HATCHING EGGS

Occasionally individuals want to mate a few turkeys for the production of hatching eggs. The turkey breeding flock is not only somewhat more difficult to manage than the chicken breeding flock but is considerably more expensive. The cost of growing the bird to maturity is several dollars (Table 18), and the amount of feed required to maintain the hens and toms during the holding and breeding periods is also substantial (Table 21).

SELECTING THE BREEDERS

Those turkeys to be kept as breeders should be started about eight months before egg production is desired. Those that come into production at too early an age lay small eggs. Fertility tends to be poorer in small eggs.

Select only the best birds as breeders. Look for candidates with good, full breasts (those with no protruding keel bones). Select for strong, straight legs, straight keel bones, and straight backs. Use only those birds that are healthy and vigorous for breeders.

MATING

One tom per twenty hens is adequate for the light-type breeders. Medium-size turkeys can be mated at the ratio of one tom per eighteen hens. Large turkeys require one tom for sixteen hens. A few extra toms should be kept to replace poor mating birds or those that die.

LIGHTING

The usual procedure is to commence lights on the toms about four to five weeks prior to mating. Toms need this light period to stimulate them to produce sperm. Hens should receive light about three weeks prior to the onset of egg production.

A thirteen- to fourteen-hour light period, artificial and natural combined, should be used. A 50-watt bulb for each 100 square feet of floor space is adequate. A time clock can be used to bracket the natural daylight hours with artificial light.

TABLE 21
FEED CONSUMPTION OF TURKEY BREEDERS
(POUNDS PER BIRD PER DAY)

TYPE OF TURKEY	HENS	TOMS
Large	.60	1.00
Medium	.45	.75
Small	.35	.65

Source: *Turkey Production AGR Handbook No. 393, USDA.*

EGG PRODUCTION

Turkeys are not as good layers as most chickens. Heavy strains may be expected to produce forty to fifty eggs during the season. Medium-size turkeys lay fifty to seventy eggs per hen, and the small strains eighty-five to one hundred eggs per bird.

First-year egg production is best. After the first year, production diminishes at the rate of approximately 20 percent each succeeding year. Egg size increases with age and hatchability tends to decrease.

MATING HABITS

Females will not mate until they commence egg production. Toms must be ready for semen production when the hen is ready to mate. The toms strut most during the mating season. When the hen is ready to mate she approaches the tom of her choice. The hen squats near the tom, the male mounts her, and copulation usually takes place. Some toms may never mount the hen, others may mount but not complete the insemination. When the male performance is not satisfactory, the hen may become disinterested and not mate for a period of time. If this happens, low fertility can result.

FERTILIZATION

Fertilization of the egg takes place in the upper part of the oviduct. Sperm are stored there for at least several days and can fertilize eggs up to three weeks after insemination. Storage life of the sperm shortens as the hens become older.

ARTIFICIAL INSEMINATION

Some breeder hens are fertilized totally by artificial insemination. Others are artificially inseminated on a supplemental basis to improve fertility where natural mating proves to be less than satisfactory.

The semen used for artificial insemination is obtained by milking the toms. Semen can be collected from the toms two or three times per week. Stimulate the tom by stroking its abdomen and pushing the tail upwards and toward the head. The male copulatory organ enlarges and partially protrudes from the vent. By gripping the rear of the copulatory organ with the thumb and forefinger from above and fully exposing the organ, squeeze the semen out with a short, sliding, downward movement. The males soon become trained and ejaculate easily when stimulated. The semen is collected in a small glass beaker or a stoppered funnel. The tom yields 2/10 to 5/10 of a cubic centimeter. The semen should be clear and free of fecal material. Some producers withhold feed and water from toms for eight to ten hours to help avoid fecal contamination of the semen.

Semen should be used within thirty minutes. About 1/40cc of semen per hen is adequate. Usually, good fertility will result when hens are inseminated two times at four-day intervals at the onset of egg production. Insemination every two or three weeks thereafter is usually desirable unless fertility is high in the mated flock.

The hen is inseminated by exposing the opening of the oviduct and inserting a small syringe without a needle into the oviduct about 1½ inches. The oviduct is exposed by pushing outward and exerting pressure on the abdomen, while at the same time forcing the tail upward toward the head. The oviduct can be exposed or protruded only in those hens that are in laying condition.

BROODINESS

Broodiness is an inherited characteristic. Turkeys are inclined to be more broody than most breeds of chickens, and some strains of turkeys are more broody than others. It is difficult to identify broody turkeys. If the birds are found on the nest in the early morning or evening it is safe to assume they are broody. Broody birds should be driven into a broody pen or yard and left there for about five days. They should be fed and watered in the broody pen. Dark spots and areas where they can rest should be eliminated, and no nests should be provided. Slat or wire floors are best for broody pens. One or more toms in the pen will tend to keep the hens on their feet and discourage broodiness. Hens usually want to mate a few days after they go broody so this practice may also improve fertility.

Broodiness can be discouraged by moving the birds to different areas, by providing roosts, and chasing birds off the nest when the eggs are gathered. Gathering eggs several times each day may also help prevent broodiness.

HOUSING

Buildings used for brooding young stock are satisfactory for breeders. Confinement is preferred to range management, where unfavorable weather is common or where predators may be a problem. Houses should be well insulated and ventilated for maximum comfort in cold or warm weather. The floor can be cement, asphalt, or wood. It should be covered with a good litter. Houses should be thoroughly cleaned and disinfected between flocks.

More floor space is required for breeders: 6 to 8 square feet per bird is usually recommended. Where only females are housed, 5 to 6 square feet is adequate for large turkeys and 4 to 5 square feet for small ones.

In warm climates or where winters are mild, breeders may be given access to fenced range or yard. They should have a shelter to protect them from bad weather and predators. The range area should be well drained and provide about 150 square feet per bird. If a yard is used, allow 4 to 5 square feet of area per bird. The shelter should be equipped with roosts and provide at least 4 square feet of floor space per bird. The feeders, waterers, and broody pen are best located inside the shelter.

NESTS

Provide one nest for every five birds. Nests should be located in a dark area of the house and be available prior to the time egg production starts so the hens will get used to them.

Some breeders prefer tie-up or trap nests, but open-type nests may be used. Nests should be 2 feet x 2 feet if they are the manger or open type. If nests are placed outside the shelter, they should be protected by a roof.

FEEDING

Put breeders on a special turkey breeder diet one month before egg production is expected. Feed according to the manufacturer's instructions. All-mash diets are preferred but mash and grain diets are satisfactory if fed in the right proportions. If too much grain is fed in relation to the mash, egg production and hatchability may be reduced. If used properly this method has the advantage of making use of home-grown grain without having to grind or mix it. Home-grown grains can be used in an all-mash program by grinding and mixing with a concentrate at home or having them mixed elsewhere.

Mash or concentrate is available in a pelleted form. There is less waste with pelleted feeds, particularly if fed on range. An insoluble grit or gravel should be available for confinement breeders. When the birds are fed and watered outdoors, move the equipment frequently to avoid muddy spots and possible sources of disease.

Six linear inches of feeder space should be provided for large turkeys and 4½ linear inches for the small birds. Provide two automatic waterers per one hundred birds. One linear inch of water space per hen is adequate. Turkeys normally consume 2 pounds of water for every pound of feed. Lack of water can be more harmful to the birds than the lack of feed.

CARE OF EGGS

Eggs should be gathered at least three times per day, more frequently if the birds tend to crowd certain nests. Frequent gathering will help to avoid breakage, dirty eggs, and possibly frozen eggs or exposure to high temperatures.

If eggs are cleaned they must be cleaned properly, using a detergent sanitizer made for egg washing. A washing temperature of 110° to 115° F. is recommended. Excessively dirty eggs gathered in hot, wet weather can easily become contaminated. Contaminated eggs often explode in the incubator.

If eggs are to be held before incubating they should be turned daily. When eggs are turned and held under good storage conditions, hatchability may be good for two weeks. Place eggs in filler flats or cartons and elevate the container at one end to slant the eggs about 30°. At least once each day, shift ends with the

Bronze Turkeys

Broad Breasted White Turkeys

Source: *Poultry Tribune*, Mount Morris, Illinois

container and elevate the opposite end. Hatchability will be improved if this is done.

FLOCK HEALTH

Hens should be debeaked when placed in the breeding pens. Remove about half of the upper beak (Figure 30). Maintain feed and water levels at a height that will permit the debeaked birds to eat and drink.

Some states require that breeders be blood tested for Salmonella pullorum and Salmonella typhoid, and possibly other diseases such as infectious sinusitus, paratyphoid, and PPLO (pleuropneumonia-like infection.) Follow the recommendations of your county agent, state Extension Poultry Specialist, or diagnostic laboratory.

CHAPTER 9

Waterfowl Production

WATERFOWL can be raised with relative ease. They are hardy and not subject to as many diseases as are other types of poultry. Numerous small flocks of waterfowl are grown to grace a small pond, as a hobby, or for exhibition. Not to be overlooked is the fact that properly finished and dressed, waterfowl are excellent for food. A number of people, in fact, prefer waterfowl to other types of poultry for the Thanksgiving and Christmas holidays. In addition, the eggs from waterfowl may be eaten and are especially good for baking. Some breeds of ducks are excellent egg producers.

DUCKS

There are several breeds of ducks available. The breed one selects will depend upon the purpose for which they are to be raised. If they are to be kept for ornamental purposes, or for hobby or exhibition purposes, the small ducks may better fit the bill. Included in this group are the White and Grey Calls, the Wood Duck, Mandarin, and Mallard.

MEAT BREEDS

If ducks are to be kept for meat purposes, the best choices include the White Pekin, the Aylesbury, the Rouen, and the Muscovy ducks.

White Pekin. The White Pekin is well suited for the production of meat. It produces good-quality meat and reaches a market weight of about 7 pounds in approximately eight weeks. It is a large, white-feathered bird. Its bill is an orange-yellow color, its legs and feet a reddish yellow color, and it has a yellow skin. The adult drake weighs 9 pounds, the adult duck 8 pounds.

The White Pekin is a fairly good egg producer. The average yearly egg

Pekin Duck

Colored Muscovy Ducks

Source: *Poultry Tribune,* Mount Morris, Illinois

production reaches approximately 160 eggs. It is a poor setter and seldom raises a brood of ducklings.

Aylesbury. The Aylesbury originated in England and, like the White Pekin duck, is a good meat bird. It reaches market weight in about eight weeks. The Aylesbury has white feathers, white skin, a flesh-colored bill, and light orange legs and feet. The eggs are tinted white. The adult drake weighs 9 pounds, the adult duck 8 pounds. It lays somewhat fewer eggs than the White Pekin and is also a poor setter.

Muscovy. There are several varieties of Muscovies, the white being the most desirable for market purposes. They produce meat of excellent quality and taste when marketed before seventeen weeks of age. They are relatively poor egg producers but good setters. The adult drakes weigh 10 pounds, the adult ducks 7 pounds. Muscovies have white skin.

Rouen. The Rouen duck reminds one of the Mallard. It has the same striking color patterns as the Mallard but is much larger. The adult drake weighs 10 pounds, the adult duck 9 pounds. Its pigmented plumage gives it a less desirable dressed appearance. It is excellent for home consumption, where the dressed appearance is not so important.

EGG-PRODUCING BREEDS

If the interest is in a breed of ducks to produce eggs, the choice should be the Khaki Campbells or the Indian Runners.

Khaki Campbell. The Khaki Campbells were developed in England. Several varieties of the Khaki Campbell ducks have been selected for high egg production. The egg production of some of these varieties is reported to average close to 365 eggs per duck in a laying year. It is interesting to note that this is a higher rate of production than the highest-producing strains of chickens.

The male has a brownish bronze lower back, tail coverts, head, and neck, and the rest of his plumage is khaki. The beak is greenish black and the legs and toes are brown. The adults weigh only 4½ pounds, thus the Khaki Campbell is not known for production of meat.

The Indian Runners. Although the Indian Runners originated in the East Indies, their egg-production qualities were developed in Europe. They are second only to the Khaki Campbell in egg production. There are three Indian Runner varieties, the White Penciled, Fawn, and White. All of these varieties have orange to reddish orange feet and shanks. The males and females weigh approximately 4½ pounds. They are not good meat birds.

BREEDS OF GEESE

The most popular breeds for meat production are the Toulouse, Emden, and African geese. Other common breeds in the United States include the Chinese, Canada, Buff, Pilgrim, Sebastipole, and Egyptian.

In choosing a breed one should consider the purpose for which they are to be raised. Geese may be raised for either meat or egg production, or as weeders, show birds, or farm pets. Some of the crossbreeds, such as the cross between the white Chinese male with the medium-size Emden female, usually result in fast-growing white geese of good market size.

Toulouse. The Toulouse has a broad, deep body and is loose feathered, a characteristic that helps give it its large appearance. The plumage is dark gray on the back, gradually shading to a light gray edge with white on the breast and white on the abdomen. The bill is pale orange and the shanks and toes are a deep reddish orange.

Emden. The Emden is a pure white goose. It is a much tighter-feathered bird than the Toulouse and, therefore, appears more erect. Egg production averages from thirty-five to forty eggs per bird. The Emden is a better setter than the

Toulouse and is one of the most popular breeds for meat. It grows rapidly and matures early.

African. The African has a distinctive knob on its head. The head is a light brown, the knob and bill are black, and the eyes dark brown. The plumage is ash brown on the wings and back and a light ash brown on the neck, breast, and underside of the body. It is a good layer, grows rapidly, and matures early. It is not as popular for meat production because of its dark beak and pin-feathers.

Chinese. The Chinese are smaller than the other standard breeds. There are two varieties, the brown and the white. Both varieties mature early and are better layers than the other breeds. They average from forty to sixty-five eggs per bird annually. The Chinese grows rapidly. It is a very attractive breed and makes a desired medium-size meat bird. It is very popular as an exhibition and ornamental breed.

Canada Goose. The Canada Goose is the common wild goose of North America. The weight of the Canada Goose ranges from about 3 pounds for the lesser Canada Goose to about 12 pounds for the giant or greater Canada Goose. The Canada is a species different from other breeds of geese. It can be kept in captivity only by close confinement, by wing clipping, or pinioning the wings. They may be kept only by permit, which must be obtained from the Fish and Wildlife Service of the United States Department of Interior in Washington, D.C.

The breed does not have the economic value of other breeds of geese—they mate only in pairs, are late maturing, and lay very few eggs.

Pilgrim. The Pilgrim is a medium-size goose and is good for meat production. The males and females of this breed can be distinguished by feather color. In day-old goslings the male is a creamy white and the female is gray. The adult male remains all white and has blue eyes. The adult female is gray and white and has dark hazel eyes.

Buff. The Buff has only fair economic qualities as a market goose and only a limited number are raised for meat. Color varies from the dark buff on the back to a light buff on the breast, and from light buff to almost white on the under part of the body.

Sebastipole. The Sebastipole is a white ornamental goose that is very attractive because of its soft plumelike feathering. The breed has long curved feathers on its back and sides and short curled feathers on the lower part of the body.

Egyptian. This long-legged but very small goose is kept primarily for ornamental or exhibition purposes. Its coloring is mostly gray and black with touches of white, reddish brown, and buff.

GETTING STARTED

Those who wish to raise a small flock of ducks or geese can best get started by purchasing day-old ducklings or goslings. This eliminates the need for keeping breeding birds and incubating the eggs. Day-old stock is available in most areas of the country. A list of those hatcheries selling waterfowl is printed in the National Poultry Improvement Plan Hatchery List. This can be obtained by writing to the United States Department of Agriculture in Washington, D.C. Other good sources of information concerning the availability of young birds include the state Extension Poultryman and the county Extension Agent in your area.

BROODING AND REARING

As is the case with other types of poultry, waterfowl need to be put on feed and water as soon as they are placed under the brooder. Young ducklings, and particularly young goslings, are somewhat hardier than chickens and turkeys. However, they do need to be kept where it is warm, dry, and free of drafts. Most any type of building that will provide these conditions is suitable for brooding ducklings or goslings. The building should be designed so that it can be ventilated without chilling the birds.

FACILITIES

The brooder house should be cleaned and disinfected well before the birds arrive. When they arrive, the house, the litter, the feeding and watering equipment, and the brooder stove should be ready. The brooder should be operating at least one day before the birds arrive. The first day, a brooder guard should be placed approximately 2 feet from the hover. It should be moved back gradually and entirely removed at about seven days. The brooder guard will confine the birds to the brooding area and prevent huddling and chilling before they are able to adjust to the location of the heat source.

Waterfowl may be started on wire, slat, or litter floors. The most practical method is to start them on litter floors. Wood shavings, sawdust, chopped straw, or peat moss are all good litter materials. The material should be free of mold, since moldy litter can cause mortality. The proper equipment arrangement for brooding is shown in Figure 10, Chapter 4.

The birds should have adequate floor space. Ducklings need ½ square foot of floor space per bird the first week, ¾ square foot the second, and 1 square foot the third week. If they are housed in confinement this space allotment will need to be increased to 2½ square feet by the time they reach seven weeks of age.

Goslings will need more floor space. They should have ½ to ¾ of a square foot of floor space the first week and 1 to 1½ square feet the second week. The amount should be increased slightly until they go on range. They can be placed on range at two to four weeks of age. No shelter is needed but some type of shade should be provided.

HEATING

Ducklings normally need supplementary heat for approximately four weeks. During warm weather or in warm climates, heat may be needed for just the first two or three weeks. Goslings usually can get by without heat after two weeks of age. Waterfowl feather very rapidly and thus do not require as long a brooding period as do baby chicks. Any brooder unit suitable for brooding chicks or turkeys is satisfactory for ducklings or goslings. Electric, gas, oil, coal, or wood-burning brooder units are all suitable.

Infrared heat lamps are excellent for brooding small groups of birds. One infrared lamp for up to thirty birds is satisfactory. To determine the number of ducklings or goslings a hover type brooder will accommodate, cut the brooder's rated chick capacity in half. They require 13 to 14 square inches of hover space per bird.

Keep the brooder temperature in the vicinity of 85° to 90° F. the first week, and reduce it approximately 5° per week during the next few weeks. The behavior of the young birds is the best guide to the temperature required (see Figure 11, Chapter 4). When the temperature is too hot, the birds will crowd away from the heat. High temperatures may result in slower weight gains and slower feathering. When the temperature is uncomfortably cold the birds will tend to huddle together under the brooder or, perhaps, crowd into corners. When a hover-type brooder is used, a night light will tend to discourage crowding; when infrared brooders are used, enough light is provided to make extra light unnecessary. When the temperature is just right the birds will be well distributed over the floor and using all the feeders and waterers.

As with other types of poultry, feather pulling can be a problem with ducks. If this vice starts, give the birds additional space. It may be necessary to debill the birds if the problem continues. This is done by nipping off the forward edge of the upper bill with an electric debeaking machine.

FEEDING AND WATERING SPACE

Keep clean drinking water in front of the birds at all times. Hand-filled water fountains or automatic waterers may be used. Place the water fountains on wire or slatted platforms or over a screened drain. This will help to keep the litter dry. Use only waterers that the birds cannot get into. Waterers should be cleaned daily.

Waterfowl should have access to range or a yard when old enough to tolerate the weather conditions. Ducklings will manage nicely on range at four weeks of age unless the weather is cold. Goslings can be placed outdoors at two weeks of age, weather permitting. They need shade and cannot tolerate chilling rains until they are well feathered on the back. When birds reach five to eight weeks of age they need shelter only during extreme weather conditions. Most commercial growers provide ponds for their birds at approximately five weeks of age; however, waterfowl can be raised without swimming water.

There are several good types of feeders for waterfowl. Feed hoppers, such as those used for chickens and shown in Chapter 2, Figures 3 through 3f, work well. Small feeders can be used until the birds are two weeks of age; large feeders should be used for older birds and breeding stock. Cover feed hoppers that are used outdoors to protect the feed from wind and rain. A range feeder designed for chickens or turkeys will work well. A plan for a range feeder is also shown in Chapter 2.

To start ducklings and goslings off it is best to use chick-box covers, shallow pans, or small chick feeders. Hanging tube-type feeders are excellent. One of these with a pan circumference of 50 inches is suitable for fifty ducklings or twenty-five goslings for the first two weeks.

Pans or troughs are satisfactory waterers. They should be equipped with wire guards or grills to prevent the birds from playing in them and spilling water. The waterers should be placed over low wire-covered frames, and preferably over drains to prevent the litter from getting too wet. A satisfactory water stand for waterfowl is shown in Figure 5, Chapter 2. Increase the size and number of waterers as the birds become older. Approximately 4 feet of trough is good for 250 young waterfowl for the first few weeks. The waterer should be wide enough to permit the bird to dip its bill and head into it.

FEEDING

Some commercial feed companies formulate rations specifically for ducks and geese. In a number of areas, however, specially prepared rations will not be available. It is possible to mix your own feed, but before doing so, write to your Extension Poultry Specialist for formulas. If feeds for ducks are not available, they may be started on crumbled or pelleted chick starter for the first two weeks. Pellets are recommended because they are easier to consume. They also reduce waste, and do not blow around like mash when used outdoors. Feed conversion is usually better with pellets; however, the lack of pelleted feed should not discourage a grower who wishes to produce ducks or geese on a small scale. Satisfactory results can be obtained with mash. If a chick starter is used, it should not contain any drugs that may be harmful to the ducklings. The use of a medicated feed is not usually necessary. Coccidiosis is not nearly the problem in ducklings that it is in chickens, though it does affect some flocks.

Ducklings and goslings do well on a starter diet that contains 22 percent protein. At two weeks of age they can be changed to a 15 to 18 percent protein diet. If a special grower diet is not available, a chick grower is satisfactory. When ducks are to be dressed as green ducks, that is at seven to nine weeks of age, this diet can be fed for the full period. In addition to a pelleted grower ration, cracked corn or other grains are often included in the diet for goslings. Feed should be kept before the birds at all times, and an insoluble grit provided. Ducks are not as good foragers as geese but may be put on pasture at four weeks of age if to be kept beyond the seven to nine week period. Feed consumption and live weight information for ducks and geese are shown in Tables 22 and 23.

THE BREEDING FLOCK

For the breeding flock, prospective breeders may be selected at about six to seven weeks of age. Choose a few extra at that time to allow for culling in the future. Birds selected for breeders should be placed on a breeder developer diet, a diet containing less energy than the starter or grower diets. Often the breeder developer diet is fed on a restricted basis to prevent the birds from becoming too

TABLE 22
AVERAGE LIVE WEIGHT, FEED CONSUMPTION, AND FEED CONVERSION RATIOS OF WHITE PEKIN DUCKLINGS AT DIFFERENT AGES /MIXED SEXES

| | | FEED CONSUMPTION | | FEED/LB. OF WEIGHTGAIN |
AGE WEEKS	LIVE WEIGHT LB.	WEEKLY LB.	CUMULATIVE LB.	TO DATE LB.
1	0.60	0.50	0.50	0.83
2	1.68	1.64	2.14	1.27
3	2.98	2.55	4.69	1.57
4	4.01	2.55	7.24	1.81
5	5.13	3.27	10.51	2.05
6	6.19	3.57	14.08	2.27
7	6.96	3.87	17.95	2.58
8	7.54	3.39	21.34	2.83

Source: Cornell University

fat. Birds that are too fat will lay fewer and smaller eggs. When feeding on a restricted basis, have plenty of feeder space or spread the pellets over a wide area to make sure that all of the birds get their share.

To obtain optimum fertility, it is best to feed a special breeder diet. Switch breeder birds to the special diet a month prior to the date of anticipated egg production. If a breeder developer diet is not available, feed a growing ration during this time.

SELECTING AND MATING THE BREEDER FLOCK

Breeder ducks are usually selected from the spring-hatched birds at six to seven weeks of age. It is possible at that time to differentiate the males from the females by their voices. The females honk and the males belch. (See the section on sex differentiation for further information.) The ratio of drakes to ducks selected is normally one to six. A few extra should be selected to take care of mortality or to permit further selection during the growing period.

Select breeders for vigor, body weight, conformation, and feathering, as well as breed characteristics. Although ducks demonstrate some selectivity in

TABLE 23
GROWTH RATE AND FEED CONSUMPTION OF
WHITE CHINESE, EMDEN GOSLINGS

AGE WEEKS	CONFINEMENT-REARED			RANGE-REARED		
	AVG. WT. PER GOSLING	CUMULATIVE FEED CONSUMPTION PER GOSLING	FEED PER LB. OF GOSLING TO DATE	AVG. WT. PER GOSLING	CUMULATIVE FEED CONSUMPTION PER GOSLING	FEED PER LB. OF GOSLING TO DATE
	LB.	LB.	LB.	LB.	LB.	LB.
3	3.3	5.25	1.75	3.3	5.25	1.75
6	8.1	17.32	2.22	7.8	12.60	1.68
9	10.8	34.92	3.32	10.1	19.97	2.03
12	12.3	47.54	3.96	11.6	30.62	2.71
14	12.8	56.09	4.48	11.8	38.10	3.31

Source: *Duck and Goose Raising,* Ministry of Agriculture and Food Ontario.

mating, they are essentially polygamous in their mating habits.

Geese are somewhat different in their mating habits. Some breeds tend to be monogamous, the Canada Goose being an example. Layer breeds mate better in pairs or trios, and the ganders of some of the smaller breeds may accept four or five females. When selecting geese for breeding flocks, select for body size, rate of growth, livability, egg production, fertility, hatchability, and breed characteristics.

HOUSING AND MANAGING THE BREEDER FLOCK

Housing for waterfowl does not have to be as tight as that for chickens and turkeys. The pen should be well lighted, well ventilated, and supplied with plenty of dry litter. Waterfowl, and especially geese, prefer to be outside even during severe winter weather. Insulated buildings are not too important. A simple shed, a small house, or an area within an existing barn can be used as a breeder pen. If more than one trio occupies a shelter, divide it to prevent fighting.

Dirt floors can be used but cement or wood floors are easier to clean and disinfect. Litter materials that are used for brooding are satisfactory for the breeders. Keep the litter dry. Wet and caked areas should be removed and dry litter added from time to time as required. Floor space requirements vary depending upon the type of bird and whether or not a yard is available. Ducks in confinement need 5 to 6 square feet per bird and 3 square feet when a yard is provided. Geese having access to a yard should have 5 square feet per bird in the house. Up to 40 square feet per bird should be provided in the yard.

Leave windows and doors open during the daylight period to allow adequate circulation of air. Good ventilation is also needed at night to prevent overheating during hot periods and to reduce condensation during the winter. Supplementary heat is not necessary.

The birds will make their own nests in the litter and lay about anywhere in the house. They prefer a somewhat secluded spot. You may provide simple nest boxes in rows along the wall. The nests should be 12 inches wide, 18 inches deep, and 12 inches high. Nests for large geese should be 24 inches square. Some prefer to not partition the nests at all—the tops and fronts of the nests are open. Straw and wood shavings make good nesting materials. The nesting material should be kept clean to prevent dirty and soiled eggs. One nest is adequate for every three to five breeders.

Pens can be kept drier if the watering devices are placed outside the house. Ducks and geese can go without water overnight, provided they do not have access to feed. When permitted to eat without available water they can choke to death on the feed. If the birds are locked in the house overnight with no water, feeders should be empty or closed. If the birds are watered in the house, again, the water supply should be placed above a screened drain.

Yards should slope gently away from the house to provide good drainage. Mud holes and stagnant puddles of water are sources of disease and parasites. Manure will accumulate in the yard, so it should be cleaned occasionally. The frequency of cleaning will depend upon the number of birds and the size of the yard. Large flocks are normally provided 75 square feet of yard space per bird.

Ducks have the tendency to run in circles if startled in the dark. This is a more serious problem with large flocks. An all-night light should be used in the house if stampeding is a problem. One 15-watt bulb for each 200 square feet of floor space is adequate to prevent this.

Breeders should not be brought into full production before they reach seven months of age. If production starts sooner there will be a problem with small eggs and low hatchability. Fourteen hours of light will stimulate them to produce in the fall and winter months when the days are short.

Females should be lighted three weeks before production, and males should be lighted four to five weeks prior to mating. Prelighting will help fertility. The lamps used for preventing stampeding in duck pens will also stimulate egg production, but larger bulbs give better light stimulation.

Ducks are much better layers than geese: they lay more eggs over a longer period of time. Geese tend to lay every other day but may lay two or more successive days. Daily gathering of eggs will help to discourage broodiness (the urge to set) and an accompanying pause in production. Separating broody birds from their mates and confining them with feed and water will sometimes discourage broodiness.

The birds should reach peak production at approximately five or six weeks after the onset of lay. The rate of production will depend upon the breed, the feeding program, and the management they receive. Older geese make better breeders than the young ones. They lay more eggs their second year, and two-year-old ganders tend to yield better fertility. Geese will lay until they are about ten years old. Ganders may be kept as breeders for five years.

EGG GATHERING AND CARE

Ducks lay most of their eggs during the night and early morning hours. The eggs should be collected early to prevent them from getting soiled and cracked. Keep the birds in the house each day until the eggs are laid. Eggs laid outside may freeze during severely cold weather, killing the embryos; they can also be badly soiled if laid outside.

Dirty eggs should be washed soon after gathering. The wash water should be warmer than the eggs: the recommended water temperature is 115° F. If cold water is used it will cause the egg contents to contract, and bacteria and other organisms may be drawn through the pores of the shell and contaminate the egg contents. This can cause the eggs to explode in the incubator. Wash the eggs in a detergent sanitizer, which may be obtained from an agricultural supply house. Wash for three minutes and follow the manufacturer's directions for use.

Badly misshaped, abnormally small or large eggs, or eggs that have been cracked should not be saved for incubation. Chances are they will not hatch. Store eggs at a temperature of 55° F. and a relative humidity of 75 percent. If hatching eggs are to be stored for more than a week prior to incubation, turn them daily. This will prevent the yolks from sticking to the shell membranes, causing a reduction in hatchability. Eggs should be stored in egg filler flats or egg cartons. Prop up one end of the container at an angle of 30° and each day change ends with the container. Eggs stored for two weeks or longer decline in hatchability quite rapidly.

INCUBATION

The incubation period for most domestic ducks is twenty-eight days. The Muscovy requires thirty-five days, the Canada and Egyptian geese thirty-five days, and all other geese thirty days.

It is possible to incubate waterfowl eggs by natural means in small flocks. A goose will cover nine or ten eggs and a duck ten to thirteen eggs. While the Muscovy duck is a good setter, most other breeds of ducks are not. If available, use a broody hen, which will cover four to six goose eggs or nine to eleven duck eggs. Turkey hens and Muscovy ducks are larger and better than chicken hens for hatching duck and goose eggs. Ten to twelve goose eggs can be set under a turkey or duck, depending upon the size of the bird. Hens should be treated for lice before the eggs are set.

Place the nest where the hen won't be disturbed during the incubation period. The nest may be placed on the ground or inside a building. Provide a source of feed and water nearby for the setting hen. If hens are used to incubate waterfowl eggs, those hatched early should be removed from the nest as soon as they are hatched and placed under a brooder. This will prevent the hen from leaving the nest before completion of the hatch or trampling some of the young if she becomes restless.

When chicken or turkey hens are used for setting, it is necessary to add moisture. Sprinkle the eggs with lukewarm water during the incubation period and place the nest and straw on the ground or on a grass-covered turf. This helps to increase the moisture. Some growers indicate that if the eggs are lightly sprinkled or dipped in lukewarm water for half a minute daily during the last half of the incubation period, they hatch much better. If ducks or geese have access to water for swimming, no additional moisture is needed. If a setting hen does not turn the eggs, mark them with a crayon or pencil and turn them daily by hand. Usually the hen will turn them.

Artificial incubation is used extensively for hatching waterfowl eggs. In the large forced-draft machines a temperature of 99¼° to 99¾° F. is usually used. For the incubation of duck eggs the wet bulb thermometer reading should be 86° to 88° F., and during the hatching period 92° F. Since the humidity requirements are so much higher for waterfowl eggs than for chicken eggs, the two should not be set in the incubator at the same time.

In the small still-air incubators the temperature should be 100½°, 101½°, 102½° and 103° F., respectively, for the first, second, third, and fourth weeks of incubation. Satisfactory results also may be obtained by running the machine at an even 102° F. for the entire period. The thermometer reading should be taken at the top of the egg. Eggs in a still-air machine are set in a horizontal position, usually, and are turned 180 degrees on each turn. Turn the eggs three to four times daily. Moisture requirements are the same for goose and duck eggs.

The eggs may be candled four to six days after incubation starts. Candling is done by passing the egg over an electric candling light in a dark room. The living embryo of the egg will appear as a dark spot at the large end of the egg near the air cell. From this dark spot radiates the blood vessels, giving a spiderlike appearance. Embryos that have died will appear as a spot stuck to the shell membrane and no radiating blood vessels will be apparent. Infertile eggs will appear clear. The infertile egg should be removed from the incubator, along with any cracked eggs that are detected.

Eggs are usually candled three days prior to hatching. At this time the fertile eggs will appear dark, except for the large air cell end. The infertiles will be clear. Discontinue turning the eggs three days prior to the end of the incubation period. From that time to hatching, do not disturb the eggs other than to open the incubator to add water.

Since the hatching period for waterfowl tends to be spread out over several hours, the large commercial hatcheries frequently take the birds out of the machines as they are hatched and dried. It is necessary to prevent chilling of the newly hatched waterfowl, however, and it may be better for the small producer to keep the incubator closed until the hatch is complete.

Some ducklings appear to need help from the shell. Conditions in small machines are sometimes difficult to control, so this practice may be justified. Birds that are helped from the shell probably should not be used for breeding stock, however, as heredity may be a factor responsible for this condition.

SEX DIFFERENTIATION OF WATERFOWL

In general, the determination of sex in ducks is somewhat easier than for geese. Drakes usually have larger bodies and heads and a soft, throaty quack, and females have a loud, distinct, harsh quack. Also, the main tail feathers are curled forward on the mature drake but not on the duck. With colored breeds of ducks, the males are more brilliantly colored than the females.

Characteristics used to distinguish between the adult male goose and the female goose are the male's longer neck and somewhat coarser and larger head, as well as his larger body. There are differences in voice between sexes, though these are somewhat difficult to determine. The African and Chinese geese have knobs on their heads, and the knobs of the male are larger than those of the female. In immature stock these differences are hard to distinguish. Even among mature geese, we find cases of oversized females and undersized males. The Pilgrim breed can be sexed from day old to maturity by plumage color: the male is light or white, and the female is a dove gray.

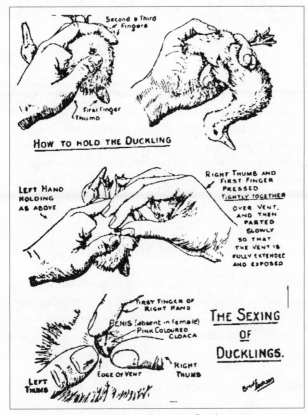

Figure 31. Sexing day-old waterfowl.

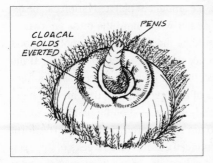

Figure 32. Appearance of the male genital organ.

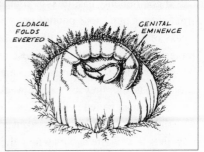

Figure 33. Appearance of the female genital organ.

SEXING GOSLINGS AND DUCKLINGS

On occasion it is desirable to be able to sex goslings and ducklings at an early age. Vent sexing of day-old ducklings and goslings is very similar. It is done in a warm room using a strong light so that reproductive organs can be seen easily. (See Figures 31, 32, and 33.)

SEXING ADULT GEESE

To sex an adult goose, hold the bird over bended knee or on a table—on its back, with the tail pointed away from you. Move the tail end of the bird out over the edge of the knee or the table so it can be readily bent downwards. Insert the index finger into the cloaca about half an inch and move it around in a circular manner several times to enlarge and relax the sphincter muscle, which closes the opening. Then apply pressure directly below and on the sides of the vent to expose the sex organs.

RANGE FOR GEESE

During warm weather, geese can be put on range as early as the second week. A good share of their feed can be forage after they are five to six weeks of age. Geese are better foragers than are ducks. They tend to pick out the young, tender grasses and clovers. They reject alfalfa and narrow-leafed tough grasses and select the more succulent clovers and grasses. Geese should not be fed wilted, poor-quality forage.

It takes approximately one acre of range to support twenty to forty geese, depending upon their size and the quality of the pasture. A 3-foot woven-wire fence will confine the geese to the grazing area.

If good-quality pasture is available, the amount of pelleted ration can be restricted to about 1 to 2 pounds per goose per week, until the birds are twelve weeks of age. The amount of feed should be increased as the supply of good forage decreases, or when the geese reduce their consumption of grass. If the birds are being grown to market as meat, they should receive pellets on a free-choice basis after they are twelve weeks of age, even though they are on range.

Geese may be fed pellets, mash, or whole grains. For the first three weeks they should receive a 20 to 22 percent protein goose starter, preferably in the form of pellets (less waste occurs with pelleted feed). After three weeks of age feed a 15 percent goose grower in the form of pellets. If a specialized grower diet or starter diet for geese is not available, the equivalent chick starter and grower can be utilized. Again, the diets need not contain coccidiostats or other medications.

Mash or whole grains can be fed alone, or they can be mixed at a 50-50 mash to grain ratio. At three weeks of age a mash to grain ratio of approximately 60-40 may be used. The proportions should be changed gradually during the growing period until, at market age, the geese are receiving a 40-60 ration of mash to grain. Depending upon the quality and quantity of available range, these ratios may be adjusted up or down slightly. For maximum growth it is important that mash and grain mixtures provide a nutrient intake of approximately 15 percent protein, the same as the all-mash diet. Insoluble grit should be available to the geese throughout the growing period.

GEESE AS WEEDERS

Geese will eat many noxious weeds and are thus used as weeders for certain crops without seriously harming the crop. Geese are good weeders in orchards and for crops such as strawberries, sugar beets, corn, cotton, and nursery stock.

For best results start the goslings as weeders at no more than six weeks of age. Provide shade and space waterers throughout the field. Keep the weeder goslings hungry. A light feeding of grain at night is enough and the amount is varied, depending upon the availability of weeds and grass in the crop being weeded.

If placed in a field before the weeds get a head start, six to eight goslings properly managed can control the weed growth on one or possibly two acres of strawberries. After the fruit begins to ripen it is best to move the goslings to other crops, or put them on range to fatten for market. A fence 30 to 36 inches high serves to confine the goslings to the area to be weeded. Any pesticide or insecticides used on adjoining tree crops must not contaminate the weeds and grasses to be eaten by the goslings.

FEATHERS

Duck and goose feathers have value. If properly cared for they may be a source of extra income, or they may be used at home. The feathers are used mainly by the bedding and clothing industries. It usually requires five ducklings or three goslings to produce 1 pound of dry feathers.

Feathers may be sold to specialized feather processing plants or a small producer can wash and dry them for home use. Wash by using a soft, lukewarm water with detergent or a little borax and washing soda. After washing, rinse, wring the moisture out, and spread the feathers out to dry.

Ducks and geese are not normally as prone to disease and parasite problems as some other types of poultry. One reason is that they are hardy birds by nature. Then, too, they are most commonly grown in small flocks and allowed to forage rather than being kept in close confinement. Where waterfowl are managed in larger numbers and under close confinement conditions, disease and parasites can be of economic consequence.

CHAPTER 10

Home Processing of Eggs & Poultry

Egg Processing

EGGS, as they come from the nest, vary in several respects. They vary in size, shape, cleanliness, shell texture, and interior quality, and some will be cracked and checked. If the right management, the right environmental conditions, and the proper feed has been provided a healthy flock, the majority of the eggs will be of good quality. If the family consumption of eggs keeps up with the birds' production, there won't be too many problems with quality differentiation or quality preservation. On occasion, however, some flocks produce more eggs than can be consumed by the family and some of them have to be sold. Eggs are sold on the basis of size and quality, and thus it is necessary to understand a little bit about egg quality determination and egg size requirements.

EGG QUALITY

The quality of an egg never improves after it is laid. Quality diminishes with time and the rate depends upon how the eggs are treated. To best understand what happens to egg quality, a knowledge of the structure and the various parts of an egg is essential.

PARTS OF EGG

There are four basic parts to an egg, namely the *shell*, the *shell membranes*, the *albumen*, and the *yolk* (Figure 34). Fastened to the shell are two membranes. It is between these two membranes that the air cell is formed, usually at the large end of the egg. When the egg is first laid this air cell is nonexistent, but with the

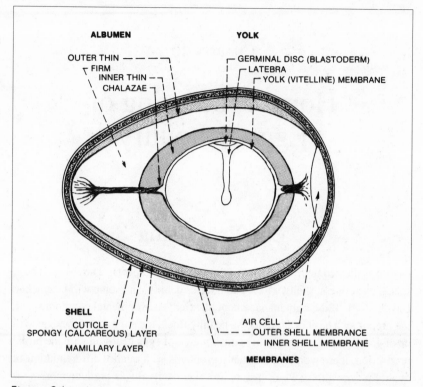

Figure 34.

cooling and contraction of the egg contents, the air cell begins to appear a short time after the egg is laid. Since the shell is porous, moisture can escape from the egg, so the air cell enlarges as the egg is held for a period of time. The rate at which the air cell increases in size depends upon age, shell texture, and the holding conditions—primarily temperature and humidity. The size of the air cell is not a factor when determining the broken-out quality of the egg, but it is a factor when candling for quality.

A relative humidity of 75 percent in the cooler will keep evaporation to a minimum. The yolk of a high-quality egg is very near the center of the egg. On the surface of the yolk is a small white spot known as the *germinal disc*, or *germ spot*. It is at this point that fertilization takes place. Whether the egg is fertile or infertile, the germ spot looks exactly the same. In other words, it is not possible to tell with the naked eye whether or not an unincubated egg is fertile. The yolk and the germ spot are enclosed by a thin membrane known as the *viteline membrane*.

The *albumen*, or white of the egg, is made up of alternate layers of thick and thin albumen. Surrounding the yolk is a thin layer of very dense albumen called the *chalazeferous layer*. From this layer of dense albumen extend the fibrous or cordlike structures called *chalazae*. These extend out towards the ends of the egg and are anchored in another layer of thick albumen. The chalazae tend to keep the yolk near the center of the egg.

INTERIOR QUALITY

Between the chalaziferous and outer layer of thick albumen there is an inner layer of thin albumen. If an egg is broken out and the thick albumen surface is cut, some of the thin albumen can be seen merging with the thick albumen.

After the egg is laid, a number of changes take place as far as the interior quality is concerned. The rate at which this takes place depends upon how the egg is handled. As mentioned earlier, when an egg is laid there is no air cell, but as the contents cool and contract the air cell develops. A further increase in the size of the air-cell is dependent upon the rate of evaporation of moisture from the egg. The rate depends upon the thickness and texture of the shell as well as the temperature, relative humidity of the storage area, and length of the storage period. Air-cell size is a factor in grading eggs, but actually has no significance as far as eating qualities are concerned.

During the holding period a change takes place in the thick albumen. It gradually breaks down into thin or watery albumen. If the egg is stored long enough, eventually there will be no evidence of thick albumen left. As a result of these changes, when eggs are broken out into a pan the contents seem to spread over a large area and the yolk has a flattened and enlarged appearance. Most individuals frown on this type of egg because it is a sign of poor quality. An egg of high quality, when broken out, will have a large amount of thick albumen adhering to the yolk. The yolk will be upstanding and practically spherical in shape. When broken out, the egg will occupy a much smaller area in the pan. If a high-quality egg that is hard-cooked is cut in two, the yolk will be well centered and the air cell small. With the low quality egg the yolk may be close to, or touching, the shell membrane, and the air cell quite large.

Egg quality can be determined relatively easily with the use of a candling light, as discussed earlier.

The standards of quality for individual shell eggs are set up by the United States Department of Agriculture (Table 24).

It should be noted that blood spots or meat spots are permissible in B-grade eggs providing the single defect, or the total of several defects, is not more than

⅛ inch. Eggs containing blood or meat spots not within these tolerances are classified as inedible eggs.

Blood spots in the egg are usually found on the surface of the yolk. They may vary in size from a small speck to a large clot with some of the blood diffused throughout the entire albumen. Blood spots are caused by the rupture of one or

TABLE 24
SUMMARY OF UNITED STATES STANDARDS FOR QUALITY
OF INDIVIDUAL SHELL EGGS

QUALITY FACTOR	SPECIFICATIONS FOR EACH QUALITY FACTOR		
	AA QUALITY	A QUALITY	B QUALITY
Shell	Clean Unbroken Practically normal	Clean Unbroken Practically normal	Clean to slightly stained Unbroken May be slightly abnormal
Air Cell	⅛ inch or less in depth May show unlimit- ed movement and be free or bubbly	3/16 inch or less in depth May show unlimit- ed movement and be free or bubbly	Over 3/16 inch in depth May show unlimited movement and may be free or bubbly
White	Clear Firm	Clear May be reasonably firm	Weak and watery Small blood or meat spots present*
Yolk	Outline slightly defined Practically free from defects	Outline may be fairly well defined Practically free from defects	Outline may be plainly visible May be enlarged and flattened May show clearly visible germ development but no blood May show other serious defects

*If they are small (aggregating not more than 1/8 inch in diameter).

Eggs that fail to meet the requirements of the above consumer grade are classified as *Restricted Eggs*, and except for "Checks" or "Cracks," cannot be sold to consumers except within the specified tolerances stated in the above grades. Dirties, leakers, inedibles, loss, checks, and incubator rejects are classified under the restricted categories.

more small blood vessels in the yolk follicle at the time of ovulation or in the oviduct during egg formation.

Meat spots are either blood spots that have changed in color due to chemical action, or tissue that has sloughed off the oviduct of the hen during egg formation. Possibly 2 percent of eggs produced contain blood or meat spots.

EXTERIOR QUALITY

In addition to the interior quality factors we have discussed, there are also exterior quality factors that must be considered. The more important exterior egg quality factors include condition or soundness of the shell and cleanliness. Leakers, dented cracks, and eggs with rough or thin shells can be detected during processing without candling. Blind checks or hairline cracks can be seen during candling. Dirty and stained eggs should not be marketed.

Weight classes of shell eggs (Table 25) are set up by the United States Department of Agriculture. The system uses the weight in ounces per dozen. For example, a single egg weighing 2 ounces is called a 24-ounce egg because a dozen of this size egg weighs 24 ounces. Uniformity of size in a market pack of eggs is important and there should be no more than a 3-ounce variation between each dozen. Small individual egg scales or larger automated scales are available for sizing eggs.

TABLE 25
UNITED STATES WEIGHT CLASSES FOR CONSUMER
GRADES OR SHELL EGGS

Size or Weight Class	Minimum Net Weight Per Dozen (Ounces)	Minimum Net Weight Per 30 Dozen Case (Pounds)	Minimum for Individual Eggs at Rate per Dozen (Ounces)
Jumbo	30	56	29
Extra Large	27	50 ½	26
Large	24	45	23
Medium	21	39 ½	20
Small	18	34	17
Peewee	15	28	—

Source: *Egg Grading Manual Agricultural Handbook No. 75.*

CARE OF EGGS ON THE FARM

As stated earlier, immediately after the egg is laid the quality begins to deteriorate. The sooner the egg is removed from the nest, cleaned, cooled, and packed, the better it is. Some management recommendations that will lead to higher-quality eggs are listed as follows:

1. Keep the birds confined.
2. Gather eggs frequently, at least three times a day.
3. If eggs must be cleaned, clean them immediately after gathering.
4. Dry-clean slightly dirty eggs.
5. Cool eggs as quickly as possible.
6. Maintain an egg storage room temperature at 50° to 55° F. and a relative humidity of 75 percent.

PROCESSING EGGS

If eggs must be washed, wash them at a temperature of 110° to 115° F. in clean water with an approved detergent sanitizer.

Washing time should be no more than three minutes. If the eggs are rinsed following the washing, use water containing a detergent sanitizer, then dry and cool. Slightly soiled eggs may be dry cleaned with an abrasive material such as emery cloth, fine sandpaper on a hand buffer, or steel wool.

On commercial egg farms the washing and sizing of eggs is done with large automated equipment. Smaller operations may use immersion washers. The eggs are immersed in warm detergent sanitizer, which is agitated to clean the eggs.

The main thing to remember is that eggs are perishable: they won't improve with age. It's important to cool them quickly and keep them cool. It's also important to realize that eggs will absorb odors and pick up off flavors. Musty odors, onions, and other vegetables can spoil the taste. Eggs should be packed in clean egg cartons, large end up, to prevent off flavors and maintain fresh quality.

A question frequently asked is, "How long can you keep eggs and still have them usable?" For years, when egg production was seasonal, eggs were stored for several months in cold storage. They weren't all AA quality when they came out of storage, but most were edible.

The shelf life of eggs can be enhanced when the shells are oil treated. A special colorless and odorless mineral oil is available in aerosol spray cans.

Usually, unwashed eggs keep better than washed eggs. It is not unreasonable to expect eggs to keep in your refrigerator for four or five weeks.

Poultry Meat Processing

AT BEST the job of processing poultry meat is a messy one. Ideally there should be two rooms available for the procedure: one for killing and picking the birds, and the other for finishing, eviscerating, and packaging. If this is not possible, the killing and plucking should be done in one operation. The room should be cleaned and the birds drawn and packaged as a second operation. This procedure will make the whole operation far more sanitary.

CARE BEFORE KILLING

Birds should be starved for about twelve hours before they are to be killed. This will give ample time for the crop and intestines to empty. Starving the birds makes the job of eviscerating much cleaner and easier. Remove them from the pen and put them into coops containing wire or slat bottoms so that they do not gain access to feed, litter, feathers, or manure.

Use care in catching and handling the birds to prevent bruising. They should be caught by the shanks and not be permitted to flop their wings against equipment or other hard surfaces. This will help prevent bruising and poor dressed appearance. After the birds are caught, keep them in a comfortable, well-ventilated place prior to killing. Overheating or lack of oxygen can cause poor bleeding, resulting in bluish, discolored carcasses.

EQUIPMENT REQUIRED

SHACKLES OR KILLING CONES

When only a few birds are to be dressed, a shackle can be made from a cord with a block of wood 2 x 2 square, attached to the lower end (Figure 35). A half-hitch is made around both legs and the bird suspended upsidedown. The block of wood prevents the cord from pulling through. Commercial dressing plants use wire shackles that hold the legs apart and make for easier plucking. Some producers make their own shackles out of heavy-gauge wire (Figure 36). Other people prefer to use killing cones, which are similar to funnels. The bird is put down through the funnel with its head protruding through the lower end. This restrains the bird and prevents some of the struggling that often leads to bruising or broken bones (Figure 36a).

KNIVES

About any type of knife is satisfactory for dressing poultry. There are specialized knives for killing, for boning, and for pinning. Six-inch boning knives work very well. For braining birds, use a thin sticking knife.

SCALDING TANK

If only a few birds are to be dressed, a 10- to 20-gallon garbage can—or any other clean container of suitable size—is satisfactory. If considerable dressing is done, a thermostatically controlled scalding vat is preferred.

THERMOMETER

Accurate temperatures are important for certain types of scalding. You should have a good, rugged dairy thermometer or some other type of floating thermometer that accurately registers temperatures between 120° to 212° F.

Figure 35.   Figure 36.   Figure 36a.

WEIGHT OR WEIGHTED BLOOD CUP

A weight or weighted blood cup attached to the bird's lower beak will prevent it from struggling and splashing blood around. The weight may be made from a window weight and attached to the lower beak by means of a sharp hook. The blood cup is not used when killing funnels are available.

KILLING

Suspend the bird by its feet with a shackle or place it in a killing cone. Hold the head with one hand and pull down for slight tension to steady the bird. With a sharp knife sever the jugular vein by cutting into the neck just in back of the mandibles. This can be done by inserting the knife into the neck close to the neck bone, turning the knife outward, and severing the jugular. It may also be done by cutting from the outside. Another method is to cut the jugular vein from inside the mouth. With this method the bird hangs with the breast toward the operator. Hold the head firmly with the thumb and first finger at the ear lobes. A slight pull with pressure will cause the beak to open. Insert the knife into the mouth, so that the point can be felt just back of the left ear lobe as you face it. (It will be opposite for left-handers). With a slight pressure and drawing outward towards the opposite corner of the mouth, cut the jugular vein at the junction of the connecting vein running across the back of the throat (Figure 37). Warning: Hold the head so that your fingers do not get in the way.

During the bleeding process restrain the birds by holding the head until the bleeding and flopping stops, or attach a weight to the lower beak. Do not grasp

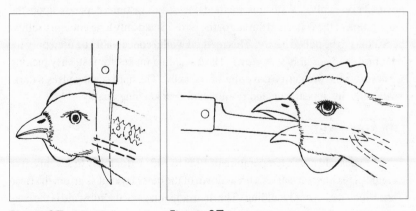

Figure 37.   Figure 37a.

the wings or legs tightly, to avoid restricting the flow of blood from these parts. A poor-appearing dressed carcass will result if the bleeding is incomplete.

DEBRAINING

Debraining loosens the feathers, making it easier to pluck the birds. Do this after the jugular vein is cut. Debraining is done when the birds are to be dry-picked, but may also be done when the birds are to be semiscalded to make the removal of feathers easier.

Insert the knife through the groove or cleft in the roof of the mouth and push through to the rear of the skull to pierce the rear lobe of the brain (Figure37a). Then give the knife a quarter-turn. This kills the bird and loosens the feathers. A characteristic squawk and shudder indicates a good stick. If the front portions of the brain are pierced it may cause the feathers to tighten. This procedure requires considerable practice before the operator becomes proficient.

PICKING

In general, there are four methods of removing feathers from birds. They are the hard scald, the subscald, the semiscald, and dry picking.

HARD SCALD OR FULL SCALD

This method is probably used more commonly on the farm than any other. The hard scald uses 160° to 180° F. temperatures for thirty to sixty seconds. After the bird is sloshed up and down in water at this temperature, the feathers are removed very easily. With this method the scalding time depends upon the temperature of the water and the age of the birds. Scald only long enough to allow the feathers to be pulled easily. This method would conceivably be used only for older birds and possibly waterfowl. Hard scalding makes for fast, easy picking but destroys the protective covering of the skin. The hard scald causes a dark, crusty, blotchy appearance and results in poorer keeping quality.

THE SUBSCALD

The subscald uses a temperature of 138° to 140° F. for thirty to seventy-five seconds. This method causes a breakdown of the outer layer of skin, but the flesh is not affected as in hard scalding. The main advantage of the subscald is the easy removal of feathers and a uniform skin color. The skin surface tends to be moist

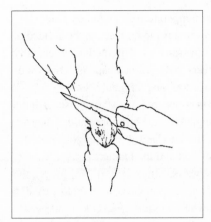

1. Make cut at least 2" long at the base of the skull.

2. Scald, follow time-temperature recommendations carefully.

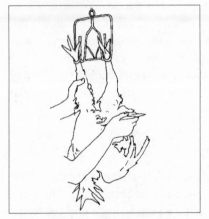

3. Use a rubbing action with the thumbs while picking.

4. Singe over flame or with blow-torch.

and sticky, however, and will discolor if not kept wet and covered. This method is frequently used for turkeys and waterfowl. The water temperature for scalding ducks is normally 135° to 145° F., and for geese from 145° to 155° F. The length of the scald is from one and one-half to three minutes.

SEMISCALD

With the semiscald the bird is sloshed up and down in water at a temperature of 125° to 130° F. Generally a water temperature of 128° to 130° F. for thirty seconds gives satisfactory results. The temperature and time varies with the age of the birds. Older birds require a higher temperature and a longer time.

Semiscalding of poultry softens and spreads the fat beneath the surface of the skin, improving the appearance of the dressed bird. This method is generally used for young birds such as broilers, fryers, roasters, and capons. Overscalding is caused by too high a temperature for too long a time.

One other technique may be utilized in connection with the semiscald or the subscald, and this is known as wax picking. After the birds are scalded they are dipped in hot wax, then removed from the wax and dipped in cold water to solidify the wax. The bird is coated with wax and the feathers and pinfeathers become imbedded in it, so that when the wax is stripped from the carcass, the feathers and pinfeathers are easily pulled out. This method is commonly used in dressing waterfowl. It is not often used when dressing chickens or turkeys.

A modification of the above-described wax-picking method is sometimes used for dressing waterfowl. Paraffin is added to the scald water to form a thick floating layer. The bird is held in the scalding vat for the required length of time and then removed slowly so that a layer of paraffin adheres to the feathers. The paraffin is given an opportunity to solidify and is then stripped off carrying with it many of the feathers and pinfeathers that are normally hard to pull.

Waterfowl feathers tend to resist thorough wetting during the scalding process. A little detergent in the scald water will help solve this problem and make the job of picking somewhat easier.

PINNING AND SINGEING

Pinfeathers, the tiny immature feathers, are best removed under a slow stream of cold tap water. Use slight pressure and a rubbing motion. Those that are difficult to get can be removed by using a pinning knife or a dull knife. Applying pressure, squeeze out the pinfeathers. The most difficult may have to be pulled.

Chickens and turkeys usually have a few hairlike feathers left following the plucking operation. These hairs can be removed by singeing with an open flame. Singeing equipment is available commercially for large operations, but for small units a small gas torch or a gas range works well. The birds are easily singed by rotating the defeathered bird in the flame.

EVISCERATING

After the carcasses are picked and singed, they should be washed in clean, cool water. As soon as they are washed they are ready to be eviscerated. Some prefer to cool the poultry before eviscerating and cutting it up because it is some-what easier and cleaner to do after it is cooled. Others eviscerate and then cool by placing the birds in ice water or cool water, which is constantly replenished. There are many methods of drawing poultry. The following methods are not the only ones but are satisfactory. Neatness and cleanliness are essential to do a satisfactory job.

The tools needed for drawing poultry are a sharp, stiff-bladed knife, a hook if the leg tendons are to be pulled, and a solid block or bench upon which to work. All equipment and working surfaces should be clean. A piece of heavy parchment paper or meat paper may be laid on the working surface and changed as necessary.

EVISCERATING ROASTING CHICKENS, TURKEYS, AND WATERFOWL

Roasters, turkeys, and capons are usually stuffed and roasted so the drawing procedure is the same for all.

It is best when eviscerating turkeys to remove the tendons from the drumsticks before removing the shanks and feet. By cutting the skin along the shank, the tendons extending through the back of the leg may be exposed and twisted out with a hook or a special tendon puller, if one is available.

The shanks and feet should be cut off straight through or slightly below the hock joint, leaving a small flap of the skin on the back of the hock joint. This will help prevent the flesh on the drumstick from drawing up and exposing the bone during the roasting process. The oil sac on the back near the tail should be cut out, as it sometimes gives a peculiar flavor to the meat. This is removed with a wedge-shaped cut.

To remove the crop, windpipe, gullet, and neck, cut the head off and slit the skin down the back of the neck. Separate the skin from the neck, and then from the gullet and windpipe. Follow the gullet to the crop and remove by cutting below the crop. The loose skin then serves as a flap that folds over the front opening and permits stuffing the bird without sewing. The neck is cut off as close to the shoulders as possible. A pair of pruning shears is handy for this purpose. The flap of skin is then folded back between the shoulders and locked in place by folding the tips of the wings over it.

Loosen the vent by cutting around it. This should be done carefully to avoid cutting into the intestine. Remove the viscera through a short horizontal cut approximately 1½ to 2 inches below the cut made around the vent. The horizontal cut should be about 3 inches long. Break the lungs, liver, and heart

Evisceration, Step By Step

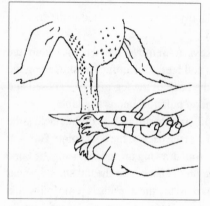

1. Use boning knife to remove feet at hock joint.

2. Cut deeply into tail and remove oil gland.

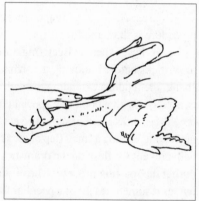

3. Cut off head before the first neck vertibra.

4. Split neck skin from shoulders to Atlas joint.

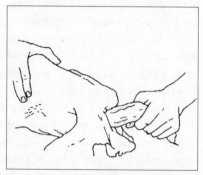

5. Pull out trachea, esophagus, and crop.

6. Cut around neck bone, double neck, and twist off.

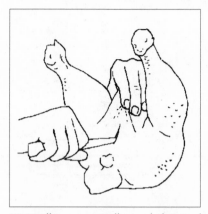

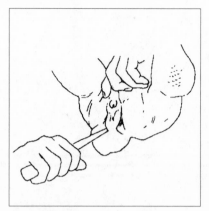

7. Midline cut: Pull up abdominal skin, then cut through skin and down towards tail.

8. Cut around vent, avoiding cut into intestines.

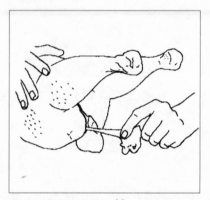

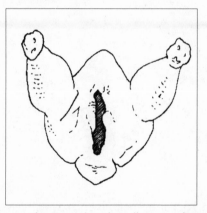

9. Remove vent and large intestine, and discard.

10. The completed midline cut for broilers and small roasters.

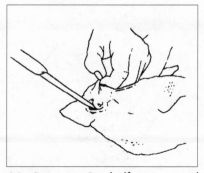

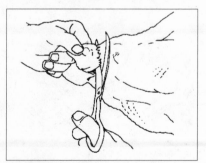

11. Bar cut: Cut halfway around vent, then...

12. Using fingers as a guide, cut with shears to free vent. Avoid large intestine.

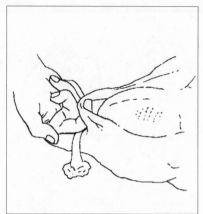

13. Pinch skin up and make 3" horizontal cut.

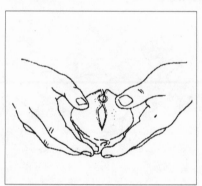

14. This leaves a "bar" of skin 1½-2" wide.

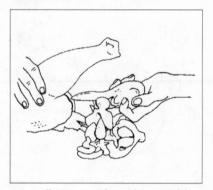

15. Pull out entrails and save giblets and liver.

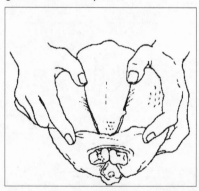

16. Cut through yellow lining of gizzard, and separate.

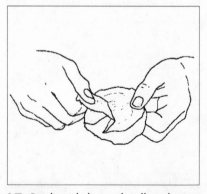

17. Peel and discard yellow lining and gizzard contents.

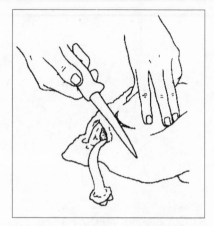

18. Tuck tail, then legs, under bar strap.

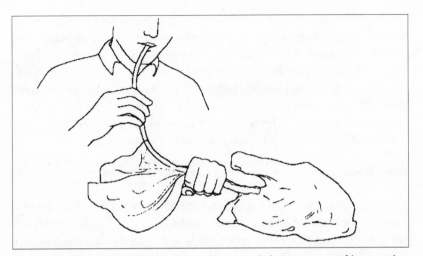

19. To air truss, place bird in plastic bag and draw air out of bag with a hose until bird is trussed.

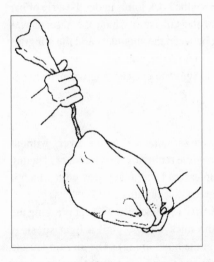

20. Twist mouth of bag four times, double and secure.

attachments carefully by inserting the fingers through the front opening. Loosen the intestines from the rear opening by working the fingers around them and breaking the tissues that hold them in the body cavity. Remove the viscera through the rear opening in one mass by inserting two fingers through the rear opening and hooking them over the gizzard, cupping the hand and using a gentle pull and slight twisting motion.

Remove the gonads (ovaries and testes). They are attached to the backbone, and can be removed quite easily by hand. Surgical capons should not have gonads unless they are "slips."

The lungs are attached to the ribs on either side of the backbone; they are pink and spongy in appearance. These can be removed by using the index finger to break the tissues attaching the lungs to the ribs. Merely insert the finger between the ribs and scrape the lungs loose.

After all the organs have been removed, wash the inside with a hose or under a faucet. Also wash the outside, removing all adhering dirt, loose skin, pinfeathers, blood, or singed hairs. Hang the birds to drain the water from the body cavity; twenty minutes should be adequate to thoroughly drain them.

TRUSSING

A properly trussed bird makes for a nicer appearance. Trussing also conserves the juices and flavors during the roasting process. One method is to place the bird on its back and draw a string across the shoulders and over the wings. Then cross over the drumsticks, bring it under the back, draw tightly, and tie at the base of the tail.

Another method of trussing is to place the hock joints under the strip or bar of skin between the vent opening and the cut from which the viscera was removed. The neck flap is drawn back between the shoulders and the wingtips are folded over the shoulders and serve to hold the skin.

The same procedures are used for killing and eviscerating waterfowl.

CLEANING THE GIBLETS

Remove the gall bladder (the green sac attached to the liver) without breaking it. If the gall bladder is broken while removing the viscera or cleaning the liver, it will likely give a bitter, unpleasant taste to any part with which it comes in contact as well as causing a green discoloration.

If the gizzard is cool and care is used, it can be cleaned without breaking the inner lining. Cut carefully through the thick muscle until a light streak is observed (do not cut into the inner sac or the inner lining). The gizzard muscle may then be pulled apart with the thumbs and the sac, and its contents removed unbroken.

EVISCERATING BROILERS AND FRYERS

The method used for drawing broilers and fryers is very similar, except that fryers are usually cut up into more pieces. Both can be drawn the same as roasters, but the usual method is to split and cut up the fryers.

The first operation is to cut off the legs at the hock joint. Remove the oil sac at the base of the tail and cut off the neck. Place the bird on its side, neck toward the operator, and the back of the bird toward the cutting hand. Split the bird with a single cut down the backbone or, if preferred, make two cuts along either side of the backbone and strip out the backbone and neck. The cuts along the backbone should be made with a stiff-bladed knife, just deep enough to sever the ribs. Cut out and remove the vent and the viscera from the bird.

To remove the breastbone, simply make a small cut in the cartilage toward the front of the breastbone. By applying a downward pressure on both sides of the bone with the thumbs and snapping it, the breastbone can be pulled out. Broilers are usually left whole or in halves, and fryers are quartered.

EVISCERATING FOWL

Fowl or old laying hens can be drawn similar to roasters and turkeys. However, they are too tough for barbecuing, roasting, or frying and are usually cut up for stewing or fricassee. One method of cutting up fowl is as follows. Remove the head, neck, crop, gullet, and windpipe as described under roasting chickens, turkeys, and capons. Next remove the feet, oil sac, wings, and legs. Make a slight cut on both sides of the body to the rear and below the breast, cutting to the backbone. Then break the fowl apart by bending back the breast. Cut away the breast, back and neck, and liver, leaving the viscera in one mass. The breast may be split into halves if desired. Separate the drumsticks and thighs by cutting through the second joint about the hock joint. Clean the giblets as described earlier.

CHILLING AND PACKAGING

It is important that the body heat be removed from the birds as soon as possible after killing. If the cooling is done slowly, bacteria can develop, causing spoilage and undesirable flavors.

If birds are to be air dried, the air temperature should be from 30° to 35° F. to cool the birds properly. The time required to cool the carcasses depends on the size of the birds and the temperature of the air. It may vary from six and one half hours for a 3-pound fowl to ten hours for a 7-pound male bird.

Water cooling can be used when air temperatures are above 35° F. If the birds have been dressed in scalding temperatures that are too high, or for too long a period of time, the air-dried birds may show a blotchy, discolored appearance of the skin. When the hard scald is used, water cooling is the preferred method

TABLE 26
PROCESSING YIELDS—POULTRY

Type	Approximate % Yield Live-to-Eviscerated
Broilers and Fryers	75
Roasters	76
Fowl—Leghorn Type	68
Fowl—Heavy Type	70
Turkeys—Heavy Roasters	80
Turkeys—Light Roasters	78
Ducks	70
Geese	68-73

Source: *New England Poultry Management and Business Analysis Manual,* Bulletin 566 (Revised).

of cooling the carcasses. Dressed birds should be cooled in pails or tanks of ice water or cold running water. The important factor is to maintain a constant temperature of 34° to 40° F. To cool the birds to an internal temperature of 36° to 40° F. requires five to ten hours in the water, depending again upon the size of the carcass.

For cut-up poultry, the birds may be cut immediately after dressing, thrown into cold water, and cooled in a much shorter time. It should again be noted that it is much easier and cleaner to draw or cut up poultry after it has been cooled.

Birds should be cooled and aged for approximately eight to ten hours. If eaten or frozen immediately after dressing, the carcasses will be tougher than if aged for a period of time.

Remove the carcasses from the running water or ice water and hang them up to dry for ten to thirty minutes before packaging. Every effort should be made to get all of the water out of the body cavity before putting it into the bag.

Wrap the giblets in a square sheet of wax paper or a sandwich-size plastic bag. Giblets can be stuffed into the body cavity. They should be wrapped well, so the carcass will not be affected if they spoil.

There are two types of bags available for poultry—one is the common plastic bag and the other is the so-called cryo-vac bag. The cryo-vac bag is a shrinkable bag that adheres closely to the bird after shrinking in boiling water. It makes a much more attractive package. It also helps to reduce the amount of water loss during the freezing process. Good plastic bags will also do a satisfactory job of maintaining quality in frozen-dressed poultry. The bags should be highly impermeable to moisture to prevent dehydration in freezer storage, which causes toughness.

Birds should be thoroughly trussed and then inserted front end first into the plastic bag. The bag should be excessively large. After the bird is in the bag, the excess air can be removed by applying a vacuum cleaner or by inserting a flexible hose into the top of the bag and applying a vacuum. Merely keep the bag snug around the hose or vacuum cleaner and suck the air out of the bag. Twist the bag several times and secure it with a wire tie or rubber band.

Fresh-dressed, ready-to-cook poultry has a shelf life of approximately five days. It should be refrigerated at a temperature of 29° to 34° F. If it is to be frozen, freeze by the third day after it is dressed and chilled. Poultry should be chilled to below 40° F. before placing in the freezer.

Weight losses from live to dressed state vary with the type, age, and size of bird (Table 26).

STATE AND FEDERAL GRADING AND INSPECTION

Processors of poultry and eggs sold off the farm are subject to the Poultry Products Inspection Act and the Egg Products Inspection Act. There are exemptions for small producers—check into them if you have questions. For information on grading and inspection programs and how they affect you, contact your state Department of Agriculture.

CHAPTER 11

Flock Health

DISEASE is not normally much of a problem with small flocks, but there are several diseases that can possibly affect your flock. It was emphasized earlier that an ounce of prevention is worth a pound of cure. If you purchase stock from a good clean source, follow a sound sanitation program, use a good feeding program, and provide a comfortable house with ample dry litter and plenty of fresh air, you will have gone a long way toward keeping your flock healthy.

Recommendations for a sound disease-control program on poultry farms should include the following:

1. Buy chicks or started pullets from disease-free sources.
2. Brood chicks away from older birds.
3. Do not mix birds from another farm with birds in your flocks.
4. Keep visitors away from your poultry buildings; do not permit them to enter pens.
5. Screen out wild birds, rodents, and predators from the buildings.
6. Clean and sanitize buildings between flocks.
7. Obtain a reliable diagnosis before starting treatment for a disease problem.
8. Never permit contaminated equipment from other poultry farms in your buildings.
9. Use a sound vaccination program.
10. Properly dispose of dead birds.

It should be realized that losses will occur in any flock. Commercial flock owners expect a monthly mortality of approximately ¾ of 1 percent. However, a disease in the flock is usually manifested by a drop in feed and water consumption, a drop in egg production, fertility, hatchability, and sick or dead birds. Birds may look dull or listless. They may show specific signs such as bloody droppings, coughing or sneezing, rattled breathing, paralysis, or other signs that distinguish them from healthy birds. When it is apparent that a disease is present in the flock, seek the advice of a trained poultry diagnostician. It is not advisable to use drugs or antibiotics indiscriminately. Sometimes this will do more harm than good and the only result may be a waste of money.

For the list of diagnosticians in your area consult the list at the back of this text. When birds are submitted to a state Diagnostic Laboratory, a sample of the flock problem should be submitted. The sample should include two or more sick birds and freshly dead birds, if any. Care should be taken to preserve dead specimens by cooling or freezing to prevent decomposition. An early diagnosis and fast treatment is always recommended for a quick solution to poultry disease problems.

There are many poultry diseases, but only the more common ones will be discussed in this text. These will not be discussed in great detail because this text is intended mainly for small-flock situations. There are many excellent texts on poultry diseases if more in-depth information is desired.

Many of the diseases described here are common to both chickens and turkeys. To avoid duplication, these will be covered in the one section. Those diseases that are peculiar to or more important in waterfowl will be covered in the section on diseases of waterfowl.

DISEASES OF POULTRY

ASPERGILLOSIS OR BROODER PNEUMONIA

Aspergillosis or brooder pneumonia is primarily a disease of young birds, though older birds sometimes have it. Symptoms include dumpy-acting birds, rapid breathing, gasping, and possibly inflamed eyes.

The disease is caused by a fungus. Fungal spores in the air are inhaled by the birds touching off an infection. Young birds may become infected in contaminated incubators or hatchers, or they may be exposed in the brooder house where they inhale the fungal spores from musty litter or feed. Older birds are sometimes contaminated by musty pen conditions or feed.

On postmortem, yellowish green nodules may appear in the lungs, trachea, bronchi, and viscera. There is no known treatment. Further spread of the infection may be prevented by culling the sick birds from the flock and cleaning and disinfecting the pens with a 1 percent solution of copper sulfate or Nystatin. Equipment should also be cleaned and disinfected. Musty litter or feed must be carefully removed to help prevent further spread of the disease.

BLACKHEAD OR HISTOMONIASIS

Blackhead is caused by a protozoan parasite. It is infrequently found in chickens but is a very common disease of turkeys of all ages. Since it can affect both chickens and turkeys, and chickens may act as an intermediate host for the organism causing blackhead, it is commonly felt to be rather risky to keep chickens and turkeys on the same farm. The term blackhead is misleading, because this symptom may or may not be present.

Cecal-worm eggs can harbor the organism over long periods of time, and when picked up by the turkeys they infect the intestines and liver. Mortality may reach 50 percent if treatment is not started and the infection checked immediately. Symptoms may be droopiness and dark heads, and brownish colored and foamy droppings. Inflammation of the intestine and ulcers on the liver may be seen upon autopsy. The incidence of the disease and its severity depends upon the management and sanitation programs used. Management factors such as sanitation of the brooding facilities, rotation of the range areas, and segregating young birds from old birds help to prevent outbreaks. Segregation of turkeys from chicken flocks is very helpful in preventing problems with blackhead.

Blackhead drugs are commonly added to turkey diets for growing birds to prevent infections. Those same drugs are also used to treat outbreak of the disease. Two of the drugs frequently used are nitarsone and furazalidone.

BLUECOMB

The cause of bluecomb is not known. It affects both chickens and turkeys, and it can hit birds at any age but usually affects chickens and pullets early in the laying cycle. Mortality can be heavy. When an outbreak occurs in laying chickens or laying turkeys, egg production is severely affected.

Characteristic symptoms include a loss of appetite, darkening of the head, and a watery diarrhea. The birds are listless and drink excessive amounts of water.

Antibiotics given in the drinking water usually result in a dramatic recovery when treatment is started early. Tetracycline, neomycin, streptomycin, penicillin, and bacitracin have all been used to treat bluecomb disease.

BUMBLEFOOT

Bumblefoot is the swelling of the foot involving the foot pad and the area around the base of the toes. It is thought to be caused by injuries resulting from bruising, a penetration of wire or thorns or other sharp objects with which the feet come in contact. Organisms such as *E. coli, Streptococcus*, and *Staphylococcus* have been isolated from the puss material found in the swollen area.

Equipment should be designed so that the birds do not get foot punctures from wire or nails. A deep, dry litter is helpful in preventing this condition, especially in those areas in front of high roosts or nests that require birds to jump to the floor.

Birds with severely swollen feet or very lame birds should be removed from the flock. Some birds may recover if the scab over the swollen area is removed and the core is squeezed out. The foot should then be washed with an antiseptic.

FOWL CHOLERA

Fowl cholera is caused by a bacterium. It is a highly infectious disease of all domestic birds including chickens, turkeys, and ducks.

The birds become sick rapidly and may die suddenly without showing external symptoms. They may appear listless, feverish, drink excessive amounts of water, and show a diarrhea.

Postmortem findings include red spots or hemorrhages on the surface of the heart, lungs, intestines, or in the fatty tissues. The birds may have swollen livers (a cooked appearance) with white spots. Treatment with sulfonamides such as sulfaquinoxalene, sulfamethazine, and others are currently recommended. Sulfaquinoxalene in the feed at the .33 percent level for fourteen days is considered to be one of the best treatments. Antibiotics are sometimes injected at high levels.

Sanitation in the poultry house, range rotation, and proper disposal of dead birds help to prevent cholera outbreaks. In problem areas, birds may need to be vaccinated. Buildings and equipment should be thoroughly cleaned and disinfected following cholera outbreak. It is advisable to vaccinate new birds coming onto farms that have had a history of cholera infection.

CHRONIC RESPIRATORY DISEASE (CRD)
MYCOPLASMA GALLISEPTICUM (MG)

Chronic respiratory disease (CRD) is an infectious and contagious disease affecting birds of all ages. It is prevalent in all areas of the country and is caused by a pleuropneumonia-like organism. It affects chickens, turkeys, and ducks.

CRD is a respiratory disease affecting the whole respiratory system, especially the accessory lungs or air sacs. It is not a killer but can cause considerable morbidity, especially if treatment is not undertaken soon and secondary invaders are not avoided. Both feed efficiency and growth rate are affected in growing flocks. Outbreaks in laying flocks cause lower feed efficiency, and peak egg production and rate of production for the laying cycle will likely be below normal.

In young birds the signs are rattling, sneezing, and sniffling; in older birds, the condition may go unnoticed. When the condition becomes complicated, there may be a nasal discharge or a foamy exudate from the eyes. There may be difficult breathing, lack of appetite, and a drop in egg production in producing birds. Young birds may be somewhat stunted and unthrifty.

Postmortem findings include cloudy air sacs with cheesy exudates and mucus in the nasal passages, trachea, and lungs.

Transmission comes through the hatching egg, through the air, and via infected equipment, feed bags, and other means.

Several antibiotics are used to treat the disease and may be administered in the drinking water, the feed, or by injection. These currently include tetracycline, tylosin, spectinomycin, and others.

Stress appears to play a key role in triggering the disease. Thus, good environmental conditions and good management are important.

Most breeder flocks in the United States are now MG negative. It is now possible to obtain MG-free stock, and the use of MG-free birds is recommended. Where several age groups are kept on farms and some flocks are MG positive, it is advisable to vaccinate young growing birds to prevent outbreaks of the disease during the laying period, with resulting drops in egg production.

INFECTIOUS BRONCHITIS

Infectious bronchitis is caused by a virus. It spreads very rapidly, hits birds of all ages, and is found in chickens. It may cause up to 50 percent mortality in young chicks and 5 to 10 percent in adult birds. It affects the reproductive organs of young birds, thus affecting future performance.

Symptoms include sneezing and coughing. Egg production drops to nearly zero in laying birds and the eggs become misshapen, soft shelled, chalky or porous, and light in color in the case of brown eggs. Interior egg quality is also affected.

Postmortem findings may be inflamed nasal passages with cheesy plugs in the lower trachea and bronchi. There is no known treatment for infectious bronchitis. Treatment with antibiotics for three to five days is usually recommended to ward off secondary invaders.

Control of infectious bronchitis is accomplished through a vaccination program. Vaccines may be administered by mass methods or individually by nasal or ocular drops. Mass methods of vaccination include drinking water, dust, or spray. The manufacturer's recommendations should be followed when using any of these products. Bronchitis vaccine is frequently given in combination with Newcastle vaccine.

COCCIDIOSIS

Coccidiosis is primarily a problem in chickens and turkeys, though on rare occasions it may be found in flocks of ducks and geese. Coccidiosis is a very common disease of poultry and is caused by a protozoan parasite, coccidia. Birds expose themselves to the disease by picking up sporulated oocysts in fecal material and litter. It should be assumed that all flocks grown on litter are exposed to the disease; birds grown on wire are not exposed to droppings and normally don't become exposed to coccidiosis.

Coccidia are host specific, that is, the coccidia that affect chickens do not affect turkeys. About nine species of coccidia affect chickens. Of these nine, three cause most of the problems. Different species affect various parts of the digestive tract. Seven species are known to infect turkeys, but only four of them commonly cause serious problems.

Ruffled feathers, unthriftiness, head drawn back into the shoulders, an appearance of being chilled, and a diarrhea that may be bloody with some forms are all symptoms of coccidiosis. If permitted to go unchecked, considerable mortality, morbidity, and poor flock results may occur.

On postmortem, lesions and hemorrhages may be seen in the various parts of the intestine, depending upon the species involved.

The disease may be prevented by feeding coccidiostats in the growing diet to permit the birds to build an immunity, or completely controlled by feeding a preventative level of coccidiostats.

Treatment involves the use of sulfonamides, amprolium, or other coccidiostats as prescribed by a diagnostician or serviceman. Sulfa compounds should be used to treat laying birds only when absolutely necessary, since they severely affect egg production.

INFECTIOUS CORYZA

Infectious coryza is a respiratory disease of chickens caused by a bacterium. It spreads rapidly from bird to bird and is thought to be triggered by stress. Mortality may be high but is usually low. Symptoms include an involvement of the sinuses and nasal passages with a nasal discharge, and swelling of the face. Egg production drops in laying flocks. Postmortem findings usually include congested nasal passages and fluid-filled tissues of the face and wattles.

Old birds from flocks that have had coryza should be treated as carriers and not mixed in with young birds. They should be disposed of or at least kept isolated from flocks that have not had the disease. Treatment with Erythromycin is usually beneficial. Some of the sulfonamides are effective but should not be used in layers. A bacterium is available for use in many states.

ERYSIPELAS

Erysipelas is mostly a disease of turkeys, caused by a bacterium. It may affect ducks and rarely chickens. Swine, sheep, and humans are also susceptible.

Symptoms are swollen snoods, bluish purple areas on the skin, and congestion of the liver and spleen. Birds may become listless, have swollen joints, and exhibit a yellowish green diarrhea. It is primarily a disease of toms because of injuries obtained during fighting. The erysipelas organism readily enters through skin breaks.

Erysipelas is considered to be a soil-borne disease, and contaminated premises are assumed to be the primary source of infection.

The disease responds to penicillin. Tetracycline is also effective. Control requires good management and sanitation.

FOWL POX

Fowl pox is found in many areas of the country. It is caused by a bacterium and spread by contact with infected birds or by such vectors as flies and mosquitoes or wild birds. There are two forms of fowl pox: the dry or skin-type and the wet or throat-type. The same organism causes both. The disease affects both chickens and turkeys.

Birds with fowl pox have poor appetites and look sick. The wet pox causes difficult breathing, a nasal or eye discharge, and yellowish, soft cankers on the mouth and tongue. With the dry pox, small grayish white lumps on the face, comb, and wattles develop. These eventually turn dark brown and become scabs.

On postmortem, cankers may appear in the membranes of the mouth, throat, and windpipe. There may be occasional lung involvement or cloudy air sacs.

There is no treatment for the disease itself, though an antibiotic may help to reduce the stress of the disease. The only means of control is to administer a vaccine. This is recommended in those areas where fowl pox is a problem.

LARYNGOTRACHEITIS

Laryngotracheitis is a viral disease that spreads very rapidly. It may be airborne, spread by contact, or by carrier birds. It affects all ages of chickens and pheasants. Its affect on birds under four weeks is mild, but may be very severe in adult birds. Mortality in some cases may be as high as 60 percent. Laryngotracheitis causes egg production to drop but has no affect on egg quality.

Signs of the disease include coughing and sneezing, and as the disease progresses, the birds find it difficult to breath. The neck is outstretched when the birds inhale, and this may be accompanied by a cawing sound. A bloody mucous may be expelled when the birds cough and sneeze.

On postmortem examination the throat, trachea, and larynx are very inflamed and swollen. There is frequently a bloody mucus or yellowish false membrane in the trachea. Severe inflammation of the larynx and trachea are in evidence and occasional cheesy plugs are found in the trachea. No treatment for laryngotracheitis is effective. In those areas where the disease is a problem, follow a vaccination program.

After an outbreak of laryngotracheitis, some birds remain carriers. These birds are a possible source of infection for nonimmune birds.

LYMPHOID LEUKOSIS

This disease has caused serious economic losses to the poultry industry for many years. Most forms of lymphoid leukosis occur in older birds or laying flocks. Mortality from this disease is seldom a problem in the young. Infected birds begin to develop tumors of the internal organs, particularly the liver, and slowly lose weight and eventually die. Acute death losses are not common with lymphoid leukosis but occur a few at a time over a long period of time.

There is no known treatment for the disease and it remains a perplexing problem for the poultry industry. Although isolation brooding and clean brooding facilities are recommended, these management practices have proved to be ineffective in controlling lymphoid leukosis.

MAREK'S DISEASE

Marek's disease is a disease of chickens caused by a herpes virus. The disease takes many forms and affects many areas of the body, including the muscle, nerves, skin, and intestines. Birds over two weeks of age may be affected. It can cause losses of 30 percent or more in pullet flocks up to the time of housing.

There is no treatment for Marek's disease, and the best control measure is vaccination of the baby chicks. This is usually done at the hatchery.

NEWCASTLE DISEASE

Newcastle disease is widespread. It is acute, highly contagious, and found in chickens, turkeys, ducks, and geese. It is a respiratory disease caused by a virus. It causes high mortality in young flocks, and production in laying birds frequently drops to zero.

Newcastle spreads rapidly through the flock causing gasping, coughing, and hoarse chirping. Water consumption increases and a loss of appetite occurs. Infected birds tend to huddle and exhibit signs of partial or complete paralysis of the legs and wings. They may hold their heads between their legs or on their backs with their necks twisted.

The disease is transmitted in many ways. It can be tracked in by people or brought in with chickens from another premise, or it can come from dirty equipment, feed bags, or by wild birds that gain entrance to the pen.

Postmortem findings may include congestion and hemorrhages in the gizzard, intestine, and proventriculas. The air sacs may be cloudy.

There is no effective treatment, though antibiotics are normally given to keep down secondary invaders. Vaccination is recommended in most areas of the country and can be administered individually or on a mass basis. Recommendations vary with the area and the type of bird. Newcastle disease vaccines can be administered intranasally, ocularly, or by wing-web. On a mass basis it can be given alone or in combination with infectious bronchitis vaccine—in the drinking water or in the form of a dust or spray. Manufacturers' recommendations should be followed for use of the product. Follow the vaccination program recommended for the area and the type of bird to be grown.

OMPHALITIS

Omphalitis is caused by a bacterial infection of the navel occurring when the navel doesn't close properly after hatching. It is found in young chicks and turkey poults, and may be caused by poor incubator or hatchery sanitation, chilling, or overheating.

Birds with omphalitis are weak and unthrifty and tend to huddle together. The abdomen may be enlarged and feel soft and mushy. The navel is infected, and the area around the navel may be a bluish black in color. Mortality may be high for the first four to five days.

There is no treatment for the disease. Most of the affected chicks die the first few days. No medication is needed for the survivors.

PARATYPHOID

Paratyphoid is an infectious disease of chickens, turkeys, ducks, and other birds and animals. It is caused by one or more of the salmonella bacteria. There are several types, but salmonella typhimurium is one of the most common in poultry and it accounts for over half of the typhoid outbreaks in poultry flocks. Transmission may come from the hen through the egg and to the chick. The organism is also found in fecal material of infected birds.

The disease is primarily one of young birds, but older birds may also be affected. In young birds mortality may run as high as 50 percent. Some birds may die without showing symptoms, while others show signs of weakness, loss of appetite, diarrhea, and pasted vents. Birds may appear chilled and huddle together for warmth. In older birds there is a loss of weight, weakness, loss of egg production, and diarrhea.

Postmortem of young birds may show unabsorbed yolk sacs, small white areas on the liver, inflammation of the intestinal tract, congestion of the lungs, and enlarged livers. Older birds usually show no lesions, though a few may show white areas on the liver.

Sulfamerazine, nitrofurans, and some of the antibiotics may reduce losses, prevent secondary invaders, and increase appetite.

Control is through sanitation and isolation of the flock from sources of infection such as wild birds, birds from other flocks, and contaminated feed and equipment.

PULLORUM

Pullorum is an infectious disease of chickens, turkeys, and some other species. It is found all over the world. Pullorum is highly fatal to birds under two weeks of age. It is caused by a bacterium, salmonella pullorum.

Mortality begins at five to seven days of age. The birds appear droopy, huddle together, act chilled, and may show diarrhea and pasting of the vent. Pullorum is sometimes called white diarrhea.

Salmonella pullorum is spread mainly from the hen to the chick through the egg. Chicks' down is a means of transmission in incubators and hatchers.

Postmortem findings in infected chicks include dead tissue in the heart, liver, lungs, and other organs, and an unabsorbed yolk sac. In adult birds there are discolored and misshapen ova; the heart muscle may be enlarged and show grayish white nodules. The liver may also be enlarged, yellowish green, and coated with exudate. Several types of blood tests help to establish a positive diagnosis for salmonella pullorum.

Sulfamethazine and sulfamerazine are effective in reducing mortality. Nitrofurans may also be effective in reducing mortality but will not cure the disease. Medication may hinder diagnosis.

Flocks that have recovered from salmonella pullorum should not be kept for replacements or breeding purposes unless they have been blood tested and found to be free of the disease.

INFECTIOUS SINUSITIS: MYCOPLASMA GALLISEPTICUM (MG)

Infectious sinusitis is a disease of turkeys caused by the same organism that causes chronic respiratory disease (CRD) in chickens.

It is transmitted through the eggs from carrier hens. As in the case of CRD, stress is thought to lower the poults' resistance to the disease.

Symptoms are nasal discharge, coughing, difficult breathing, foamy secretion in the eyes, and swollen sinuses, accompanied by a drop in feed consumption and loss of body weight. Air sac involvement may be in evidence on postmortem.

Antibiotics and antibiotic vitamin mixtures will help. Chlortetracycline or oxytetracycline in the feed is beneficial. Individual bird treatment with injectable penicillin or streptomycin in the sinuses is recommended.

INFECTIOUS SYNOVITIS: MYCOPLASMA SYNOVIAE (MS)

Synovitis is an infectious disease of chickens and turkeys caused by a pleuropneumonia-like organism (PPLO). At first it was identified as a cause of infections of the joints, but more recently it has been identified with a respiratory disease. The disease can affect birds of all ages. It is found in all areas of the United States and is a serious problem, particularly in broiler flocks.

Symptoms include lameness, hesitancy to move, swollen joints and foot pads, weight loss, and breast blisters. Some flocks show signs of the respiratory problem. Dying birds may show a greenish diarrhea.

Upon postmortem examination, swelling of the joints and a yellow exudate—especially in the hock, wing, and foot joints—is in evidence. Internally, the birds may show signs of dehydration and enlarged livers and spleens. Birds showing the respiratory involvement are not easy to spot. Postmortem examination may show the air sacs filled with liquid exudates.

The most common means of transmission is through infected breeders. Poor sanitation and management practices also contribute. In most primary breeding flocks of chickens, the infection has been eliminated by serological testing. Testing hasn't been as successful in turkey breeding flocks. Treatment is prescribed for market egg flocks, but breeder flocks with infectious synovitis should probably be disposed of—it can be passed on to their offspring. Antibiotics yield some results. These should be given by injection or in the drinking water; some prefer to use both simultaneously for best results.

FOWL TYPHOID

Fowl typhoid is caused by a bacterium, salmonella gallinarum. It affects chickens, turkeys, ducks, and other species of birds. The disease may be present wherever poultry is grown.

Affected birds may look ruffled, droopy, and unthrifty. Other symptoms may include pale combs and wattles, loss of appetite, increased thirst, and a yellowish green diarrhea.

On postmortem, the liver may have a mahogany color. The spleen may be enlarged and there may be pinpoint necrosis in the liver and other organs. There may be pinpoint hemorrhages in the fat and muscle tissue and evidence of enteritis.

Prevention, treatment, and control are much the same as for pullorum disease, except there is a vaccine available that is useful in controlling mortality. The blood test used to detect pullorum disease will also detect fowl typhoid in carrier birds.

DISEASES OF WATERFOWL

Waterfowl raised in small flocks very seldom have disease problems. Where waterfowl are kept in large numbers and concentrated in relatively small areas, disease may be more prevalent. Some breeds of waterfowl exhibit more resistance to diseases than others. For example, the Muscovy duck appears to be more resistant to diseases common in the Pekin and Runner.

Only the more common diseases of waterfowl will be considered in this section. It should again be emphasized that when disease problems occur, it is best to get a reliable diagnosis from a poultry diagnostician.

AMYLOIDOSIS

This is one of the common diseases of adult ducks and geese. Amyloidosis is also called wooden liver disease and can be quite readily recognized by the hardness of the liver. Mortality may reach as high as 10 percent in some flocks. The cause of the disease is not known and there is no known cure. Postmortem may show large accumulations of fluids within the body cavity.

ASPERGILLOSIS OR BROODER PNEUMONIA

This disease was discussed earlier under chicken and turkey diseases. The causes and control measures are the same for waterfowl.

BOTULISM

Botulism occurs in both young and adult waterfowl and in other types of poultry. It is caused by the bacterium known as *clostridium botulinum*. This organism grows in decaying plant and animal material. Birds feeding on material containing the toxins produced by the bacteria lose control of their neck muscles. For this reason it is sometimes referred to as "limber neck." Infected waterfowl may drown if swimming water is available to them. Maintaining dry, clean facilities and removing decaying vegetation or spoiled feed will usually prevent this disease. A mixture of Epsom salts, at the rate of 1 pound per 5 gallons of water, has been reported as an effective treatment.

COCCIDIOSIS

As mentioned earlier, coccidiosis is a very infrequent problem with ducks. The type of bacteria causing the disease in ducks is different from those causing the problem in chickens and turkeys, but control and treatment is the same as for other species of birds.

DUCK VIRUS ENTERITIS OR DUCK PLAGUE

Duck plague affects ducks, geese, swans, and other aquatic birds. It is caused by a virus that is transmittable by contact. It commonly occurs where water is available for swimming, and has affected ducks in commercial flocks in the eastern United States.

Signs of the disease are watery diarrhea, a nasal discharge, and general droopiness. The symptoms develop three to seven days after exposure. Death frequently results in three or four days. Generalized hemorrhages in the body organs may be seen on postmortem.

There is no satisfactory treatment. Only strict sanitation and rearing ducks in pens with access to drinking water will reduce and help control the disease.

Exposure of domestic ducks to duck virus enteritis from migratory waterfowl can be avoided by keeping the susceptible ducks in houses or wire enclosures.

FOWL CHOLERA

Fowl cholera has been discussed earlier under chicken and turkey diseases. The disease also attacks ducks and geese. Control and treatment of fowl cholera in waterfowl is the same as for other birds.

KEEL DISEASE

Keel disease occurs in young ducklings the first few days after hatching. Affected birds appear thin and dehydrated.

Fumigation of the hatching eggs and a thorough washing and fumigation of the hatcher between hatches are recommended to reduce the number of bacteria to which the young ducklings are exposed. Clean, warm brooding facilities and good feed and fresh water will also help control the disease.

NEW DUCK DISEASE OR INFECTIOUS SEROSITIS

Caused by a bacterium, new duck disease is one of the most serious diseases affecting ducklings. The symptoms resemble those of chronic respiratory disease of chickens. The first signs of the disease are sneezing and loss of balance. Losses of up to 75 percent have been recorded. Death often is due to water starvation rather than to the primary infection. Antibiotics and sulfa drugs have been used with some success.

VIRUS HEPATITIS

Virus hepatitis outbreaks can cause 80 to 90 percent mortality in flocks of young ducklings. It is a highly contagious disease and strikes ducklings from one to five weeks of age.

Vaccination of the female breeding stock will prevent this disease. Antibodies produced by the vaccinated laying ducks are passed on through the egg to the young ducklings. It is recommended that ducklings be brooded and reared in strict isolation, especially for the first five weeks, to help prevent outbreaks of the disease.

PARASITES OF POULTRY

Although there are many parasites of poultry, relatively few of them are of major importance. Some of the parasites live inside the bird and others live on the outside. Certain external parasites affect growth and egg production. Some of the internal parasites can cause setbacks in weight gain, loss of egg production, and even death if the infestation is severe. A few of the intestinal parasites harbor other disease organisms harmful to chickens and turkeys.

INTERNAL PARASITES

Large Roundworm. A light infestation of roundworm probably does little damage to the bird. When worms become numerous, however, the bird can become unthrifty and feed conversion can suffer. The worms themselves seldom cause mortality, but when present with other diseases they may cause an increase in mortality.

The large roundworm is 1½ to 3 inches long. It is found in the middle of the small intestine and causes setbacks in weight gain and a loss of egg production in adult birds. The piperazine compounds are used to treat roundworms, and can be given in the water, in the feed, or in capsule form.

Birds infect themselves with the roundworm by picking up the egg from feces, the litter, or other infested materials. It is therefore desirable to keep the litter as clean as possible to prevent the birds from picking up the eggs. Birds on wire floors or in cages do not have worms.

Capillaria Worm. The capillaria worm is a very small worm ½ to 1½ inches in length. It inhabits the upper part of the intestine, including the esophagus, crop, and occasionally the ceca. It can cause inflammation of the intestines, diarrhea, weakness, and in laying birds, lowered egg production. It is very difficult to see capillaria worms with the naked eye, and so sometimes their

presence is overlooked on autopsy. Capillaria worm infestations prevent efficient utilization of vitamin A in the bird; treatment rations are often supplemented with additional vitamin A. Hygromycin added to the diet has been
reported to be helpful in some cases.

Capillaria has the same direct life cycle as the roundworm. The embryo-
nated egg is picked up from the litter by the bird. The earthworm is also an intermediate host. Sanitary conditions are very important for the control of capillaria
worms. It is very difficult to get rid of them where dirt floors are used in poultry
buildings.

Cecal Worm. The life cycle of the cecal worm is similar to that of the large
roundworm. These worms are found in the ceca instead of the small intestine.
Their length is ½ to ¾ inches. By itself it probably has little economic
significance in chickens, but its eggs harbor the organism that causes blackhead
disease of turkeys. In severe infestations, cecal worms may contribute to a
somewhat unthrifty condition, weakness, and emaciation. It is rather hard to treat
because most of the intestinal contents bypass the ceca during the process of
digestion. Phenothiazine or hygromycin B have been used successfully as
treatments.

Gapeworm. The gapeworm attacks the bronchi, trachea, and lungs. It can cause
pneumonia, gasping for breath, or even suffocation. High mortality may occur
in young birds, especially turkeys and pheasants.

The use of thiabendazole in the feed for two weeks is effective in eliminating
gapeworms. Tetramisole has been reported as being an effective treatment when
given in the drinking water for three days.

Gizzard Worm. The gizzard worm is reddish in color, about ½ to 1 inch in
length. It is found under the lining of the gizzard, and it impairs digestion due
to the damage to the gizzard. No treatment is known, but the disease may be
controlled by avoiding exposure of the birds to such intermediate hosts as
grasshoppers, beetles, weevils, and other insects.

Tapeworm. There are several species of tapeworms varying in size from microscopic to 6 to 7 inches long. They are flat, white, and segmented. They inhabit
the small intestine and cause a loss of weight and a loss of egg production in adult
laying birds. Treatment recommended for tapeworms is dibutyltin dilaurate.
This drug is sold in combination with piperazine and is effective in the removal
of tapeworms. One caution should be observed in the use of this material. It

should be used only for severely infected flocks and should not be fed to laying birds.

The effective control of internal parasites depends primarily upon a program of cleanliness and sanitation. Parasitic eggs can remain viable in the soil for more than a year. This makes it important for producers to practice rotation of poultry runs or yards and if possible, to keep them clean and under cultivation to prevent the birds from picking up the parasites.

EXTERNAL PARASITES

Lice. Lice are chewing or biting insects that cause birds considerable grief. With severe infestations, egg production, growth, and feed efficiency can suffer. They cause an irritation of the skin with a scab formation.

Lice spend their entire lives on the birds; they will die in a matter of a few hours if they leave. The eggs are laid on the feathers, where they are held with a gluelike substance. The eggs hatch in a few days to two weeks. Lice live on the scales of the skin and feathers.

There are several types of lice that attack poultry, including the body louse, shaft louse, wing louse, large chicken louse, and the brown chicken louse. The body louse is one of the most common lice of poultry and usually affects older birds. The lice and its nits (eggs) are seen on the fluff, the breast, under the wings, and on the back. The head louse attacks the feathers of the head and seems to be most harmful to young birds.

Materials that may currently be used to treat lice are carbaryl, malathion, and co-ral. These should be used according to the directions on the label. Birds should be examined frequently for signs of a reinfestation.

Northern Fowl Mite. Northern fowl mites are a reddish dark brown in color. They are found around the vent, the tail, and the breast of the birds. These mites live on the birds at all times and may be seen on eggs in the nest when there is a severe infestation. They attack the feathers, suck blood, and cause anemia, loss of weight, and loss of egg production. Materials recommended for treatment of the northern fowl mite include carbaryl, co-ral, and permethrin. Use according to directions.

Red Mite. Red mites are not found on the birds during the daytime—they feed during the nighttime. They may be seen on the underside of roosts, cracks on the wall, or seams of the nest during the daylight hours. Other signs are salt-and-pepper-like trails under roost perches or clumps of manure on the furniture. Red mites are blood suckers. They cause irritation, loss of weight, loss of egg production, and anemia. Laying birds may refuse to use infested nests. Use carbaryl (sevin), coumaphos, or malathion according to manufacturer's directions.

Depluming Mite. The depluming mite burrows into the skin and causes an irritation at the base of the feather. In an attempt to relieve the itching, the birds pull out feathers until they are nearly naked in cases of severe infestation.

To control the depluming mite, use a dip containing 2 ounces of sulfur and 1 ounce of soap per gallon of water. The birds should be dipped to wet the feathers to the skin. Dip only on warm days. The treatment should be repeated in three or four weeks if necessary.

Tropical Chicken Flea. This flea clusters on the comb, wattles, and earlobes and around the eyes of the bird. It causes irritation to the membranes of the eye and may cause eventual blindness and death. It also causes a loss of weight and lowered egg production. Use carbaryl, coumaphos, or malathion according to the manufacturer's directions.

Fowl Tick and Blue Bug. The fowl tick is a flat, egg-shaped, reddish brown bug. It resembles the ticks. The adults are nocturnal feeders. They cause a loss of appetite, a loss of body weight, and lowered production. Loss of blood may result in anemia. The ticks may also carry diseases: heavy infestation may lead to so-called tick paralysis. Carbaryl, coumaphos, and malathion are effective materials for the control of ticks and blue bugs.

Bed Bug. The bed bug is a brown or yellow-red bug found under roosts and in cracks and crevices. The bugs are nocturnal feeders; they do not remain on the birds during the daylight hours. Birds pull out their own feathers in an effort to relieve the intense itching caused by bed bugs. The treatment for this parasite is the same as for the northern fowl mite.

Scaly Leg Mite. This mite is microscopic in size. It burrows into the skin of the legs and under the scales, and may occasionally attack the comb and wattles. It may cause the birds to go lame or be crippled for life. There is evidence of scales and crusts on the legs, and the legs may be swollen. Scaly leg mite is not common where proper sanitation is practiced.

Treatment is on an individual-bird basis. Dip legs in a mixture of one part kerosene plus two parts raw linseed oil, mineral oil, or warm Vaseline.

Chiggers. The chigger that attacks poultry is the same red mite that attacks humans. Birds infested with chiggers may become droopy and emaciated. Chiggers attach themselves to the birds' skin in clusters under the wings and on the back and neck. They may cause abscesses and inflamed areas. Birds may even die with severe infestations.

One aid for controlling chiggers is to keep the grass and weeds on range or in the yards cut short. The range may also be dusted with 1 percent malathion at the rate of 40 pounds per acre.

PRECAUTIONS FOR DRUG AND PESTICIDE USE

Use drugs and pesticides with caution. To get optimum production results from poultry, it's sometimes necessary to use drugs or pesticides. Before they are used, however, make certain of the problem and then use the material of choice according to recommendations. Never use a material that is not registered for use on poultry.

Neither drugs nor pesticides are intended as substitutes for good preventive management. They work best when used in combination with good sanitation and sound management practices. Early diagnosis and treatment of disease or parasite problems is recommended.

Agencies of the federal government have set up definite withdrawal periods and tolerance levels for various materials used in poultry. The Food and Drug Administration, for example, has set minimum periods of time when certain drugs must be withdrawn prior to poultry slaughter. These withdrawal periods vary from one to two days up to several days, and are subject to change from time to time. Maximum tolerances for residues of certain materials are also established. Some insecticides can be used around poultry but not directly on the birds or on the eggs or in the nests. Some pesticide materials cannot be used within a certain number of days prior to slaughter. Frequency of use is also an important consideration with some materials.

Withdrawal periods and tolerances, as mentioned above, do change, and accepted treatment materials change, so the specific precaution for use of various materials will not be dealt with here. It is hoped that the point has been strongly made that drugs and insecticides should be used discriminately, following all the precautions on the label for the use of a given material. Improperly used, they can be injurious to humans, animals, and plants.

Store all drugs and pesticides in the original containers in a locked storage area. Keep them out of the reach of children and animals. Avoid prolonged inhalation of sprays or dusts, and wear protective clothing and equipment where recommended. Be safe!

NUTRITIONAL DISEASES

A number of poultry diseases are caused by nutritional deficiencies or imbalances. With the well-formulated diets on the market today, nutritional problems are uncommon. Therefore, a thorough discussion of nutritional diseases will not be undertaken. To point out the importance of good nutrition, however, I'll just mention a few of the more common nutritional diseases.

RICKETS

Rickets is caused by a deficiency of vitamin D_3. Birds with rickets appear to be weak, have stiff swollen joints, soft beaks, soft bones of the leg, and enlarged rib bones. A deficiency of phosphorus can also cause rickets.

PEROSIS

Perosis is a leg weakness usually caused by a deficiency of manganese in the diet. Choline, niacin, and biotin may also have a role. With this deficiency, most chicks will respond to manganese fortification in the feed.

ENCEPHALOMALACIA

Encephalomalacia, or vitamin E deficiency, is commonly called "crazy chick disease" because it causes staggering, uncoordination, and paralysis of affected chicks. Vitamin E deficiency causes a part of the brain to degenerate. Positive diagnosis of this disease requires a study of the brain tissues.

CURLED TOE PARALYSIS

Curled toe paralysis is caused by a riboflavin deficiency in the feed. Feed conversion, growth rate, and bird mortality are adversely affected. One or both feet may be affected. Additional fortification of the feed with riboflavin will help to prevent further development of this disorder.

NUTRITIONAL ROUP—VITAMIN A DEFICIENCY

Birds deficient in vitamin A become very unthrifty. Exudates may form in the eyes and nostrils and whitish pustules develop in the mouth, esophagus, and crop. Mortality in young poultry can be very high if the deficiency is not corrected. Adult birds exhibit reduced production, lowered hatchability, and decreased livability in the offspring. Yellow corn, legumes, grasses, and fish oils are good sources of vitamin A.

PREDATORS

Rats and mice are a common problem on poultry farms. They like to eat poultry feed. They may destroy feed bags if they are hung up where they can get at them. They destroy insulation in poultry buildings and are carriers of disease. A few rats or mice will quickly produce a large population if not controlled.

Rats and mice like hiding places. Given places to hide in or around the chicken house, they will set up housekeeping. They like to inhabit the space between double-sheathed walls, stone foundations, under floors, and in and around junk. Good housekeeping will help to eliminate hiding spots and prevent rodent problems.

Trapping can be helpful if the populations of rats and mice are small. Keeping a cat in the area will also help to control them. Baiting is the best way of handling large populations.

Anticoagulant baits are the most effective means of eradicating rats and mice. Anticoagulants cause rodents to bleed to death internally after they have consumed the bait for several days.

To be effective, bait stations (Figure 38) need to be set up near walls, burrows, and paths that the rats and mice tend to follow. These bait stations need to be kept full of fresh bait—but no more than they will eat in two or three days. If bait gets old and musty they will not touch it. Place fresh water near the baiting stations to increase consumption of the bait.

Ideally, grain should be stored in rat- and mouse-proof bins or containers. Covered galvanized garbage cans are excellent for storing feed for small flocks.

OTHER PREDATORS

Although rats and mice are the biggest nuisance to small flocks, one should be aware of other possible problems.

At times, dogs and cats may raid the poultry yard or house. Cats can be a problem if they gain access to a pen of young chicks. Occasionally, dogs get into the poultry house or poultry yard and manage to kill young or adult birds. Usually, the real loss comes from piling of the birds and subsequent suffocation due to fright.

Foxes, raccoons, skunks, weasels, and mink may frequent areas where poultry is kept. The work of a fox can be identified by the signs of hair on the fence, or a hole where he crawled through. He usually kills several birds, and frequently leaves them behind, partially buried. When the raccoon comes to visit he tends to eat the crops out of the birds, and some heads may be missing. Given the opportunity, he returns every fourth or fifth night to get more.

Owls and hawks also like chicken. If one or two birds are killed every night and the heads and necks are missing, an owl may be the problem.

When birds are killed occasionally and are badly beaten, a dog problem is indicated. If several birds are killed on an irregular basis, it may also be the work of a mink or weasel, in which case there will be small bites around the neck and head.

Confinement rearing has prevented many losses from predators. When birds are ranged or yarded they should be closed in a shelter at night. If range shelters are used, screen the space beneath the shelter. Installation of an electric fence a few inches above the ground may serve to discourage some predators.

DEAD BIRD DISPOSAL

Disposal of dead birds in a sanitary manner reduces the spread of disease as well as fly and odor problems. Dead birds can be incinerated, placed in disposal pits, or buried. Those flock owners who are more fortunate may have a pickup service available.

INCINERATION

A good incinerator is an excellent means of dead bird disposal. There may be objectionable fumes to contend with, especially if there are neighbors nearby. Modern incinerators are quite efficient and most have afterburners, which tend to reduce the odors. The main disadvantages are the necessary investment and the high operating costs, particularly for small-flock use.

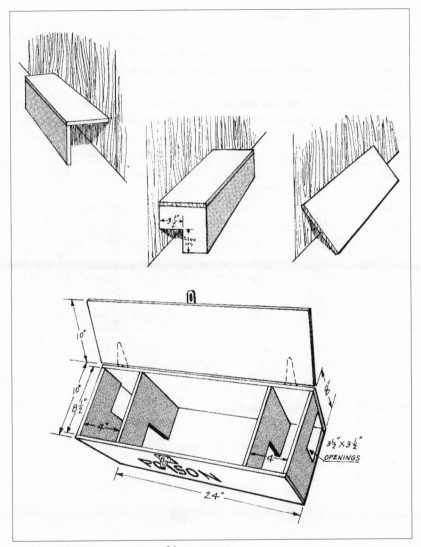

Figure 38. Various types of bait stations.

DISPOSAL PIT

A properly constructed disposal pit is an efficient and practical way to dispose of dead birds. They are relatively inexpensive to build and require little maintenance. The pit must be constructed in well-drained soils at least 100 feet away from wells or springs, and where drainage is away from other water supplies in the neighborhood.

The disposal pit (Figure 39) is capable of handling the mortality from flocks of about one thousand birds for three years. For smaller flocks, the pit size can be scaled down somewhat. Make sure the use of a disposal pit is permitted in your area.

Deep burial is a practical and satisfactory method of dead bird disposal for small flocks. It must be done properly, however. If buried too shallow, the carcasses may be dug up by dogs and other animals.

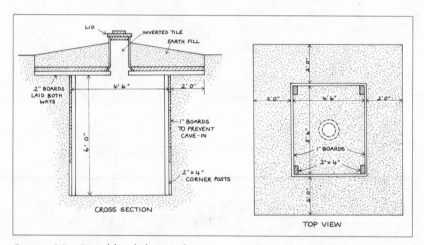

Figure 39. Dead bird disposal pit.

Appendices

Diagnostic Laboratories by State

Alabama
C.S. Roberts Veterinary Diagnostic
  Laboratory
P.O. Box 2209, Wire Road
Auburn, AL 36831

Poultry Producers Diagnostic
  Laboratory
1724A 2nd Avenue, N.W.
Cullman, AL 35055

Alaska
Cooperative Extension Service
University of Alaska
Fairbanks, AK 99709

Arizona
Arizona Veterinary Diagnostic
  Laboratory
University of Arizona
Tucson, AZ 85721

Arkansas
Arkansas Livestock Poultry
  Commission Diagnostic
  Laboratory
P.O. Box 5497
Little Rock, AR 72215

California
Department of Food and Agriculture
W. Health Sciences Drive
Davis, CA 95616

California Veterinary Diagnostic
  Laboratory
3290 Meadowview Road
Sacramento, CA 95832

San Diego Veterinary Laboratory
  Bldg. 4
5555 Overland Avenue
San Diego, CA 92123

Connecticut
Department of Pathology
University of Connecticut
Storrs, CT 06268

Delaware
University of Delaware
Poultry Diagnostic Laboratory
RR2, Box 47
Georgetown, DE 19947

Delaware (continued)
USDA ARS Poultry Research
  Laboratory
RD 2, Box 600
Georgetown, DE 19947

Sterwin Laboratories, Inc.
Millsboro, DE 19966

Florida
Bureau of Diagnostic Laboratories
P.O. Box 1031
Dade City, FL 33525

Bureau of Diagnostic Laboratories
P.O. Box 460
Kissimmee, FL 32742

Live Oak Diagnostic Laboratory
P.O. Drawer O
Live Oak, FL 32060

Florida Dept. of Agriculture &
  Consumer Services
8701 N.W. 58th Street
Miama, FL 33178

Miami Spring Diagnostic Laboratory
7614 S.W. 129 Court
Miami, FL 33183

Georgia
Diagnostic Assistance Laboratory
University of Georgia
Athens, GA 30602

Georgia Poultry Laboratory
Veterinary Poultry Diagnostic
  Laboratory
P.O. Box 349

Canton, GA 30114

Georgia Poultry Laboratory
1126 Lamar Street
Dalton, GA 30720

Georgia Poultry Laboratory
RR 5, Box 3
Douglas, GA 31533

Georgia Poultry Laboratory
Oakwood Road, P.O. Box 20
Oakwood, GA 30566

Diagnostic and Investigational
  Laboratories
P.O. Box 1389
Tifton, GA 31793

Hawaii
Veterinary Laboratory Branch
Hawaii Dept. of Agriculture
99-762 Moanalua Road
Aiea, HI 96701

Illinois
Animal Disease Laboratory
Shattuc Road
Centralia, IL 62801

Animal Disease Laboratory
1855 Windish Drive
Galesburg, IL 61402

Indiana
Animal Disease Diagnostic
  Laboratory
RR 1
Dubois, IN 47527

Animal Disease Diagnostic
  Laboratory
Purdue University
West Lafayette, IN 47907

Iowa
Veterinary Diagnostic Laboratory
Iowa State University
Ames, IA 50011

Hy-Line International Customer
  Service Lab
1915 Sugar Grove
Dallas Center, IA 50063

Kansas
Veterinary Diagnostic Lab
College of Veterinary Medicine
Manhattan, KS   66502

Kentucky
Department of Veterinary Science
  Diagnostic Laboratory
University of Kentucky
Lexington, KY 40506

Louisiana
Louisiana Veterinary Medical
  Diagnostic Laboratory
P.O. Box 25070
Baton Rouge, LA 70894

Northwest Louisiana Livestock
  Diagnostic and Research Laboratory
P.O. Box 2156
Natchitoches, LA 71457

Maine
University of Maine Pathology
  Diagnostic and Research Lab

Hitchner Hall
Orono, ME 04469

Maryland
Maryland Dept. of Agriculture
Animal Health Laboratory
R 1, Box 145
Centerville, MD 21617

Maryland Dept. of Agriculture
Animal Health Diagnostic Laboratory
4901 Calvert Road
College Park, MD 20740

Maryland Dept. of Agriculture
Animal Health Laboratory
P.O. Box 1234
Frederick, MD 21701

Maryland Dept. of Agriculture
Animal Health Lab
P.O. Box 376
Oakland, MD 21550

Maryland Dept. of Agriculture
Animal Health Lab
P.O. Box J
Salisbury, MD 21801

Michigan
Animal Health Diagnostic Laboratory
P.O. Box 30076
Lansing, MI 48909

Minnesota
Veterinary Diagnostic Laboratory
E 220 Diagnostic Research Building
College of Veterinary Medicine
St. Paul, MN 55108

Mississippi
Central Laboratory
P.O. Box 1510
Forest, MS 39074

Mississippi Veterinary Diagnostic
  Laboratory
P.O. Box 4389
Jackson, MS 39216

Missouri
Veterinary Medical Diagnostic
  Laboratory
University of Missouri
Columbia, MO 65211

Missouri Dept. of Health Labs
307 W. McCarty
Jefferson City, MO 65101

Missouri Dept. of Agriculture
Veterinary Diagnostic Laboratory
1922 N. Broadway
Springfield, MO 65803

Montana
Montana Veterinary Diagnostic
  Laboratory
P.O. Box 997
Bozeman, MT 59771

Nevada
Nevada Dept. of Agriculture
Animal Disease Laboratory
P.O. Box 11100, Capitol Hill Avenue
Reno, NV 89510

New Hampshire
University of New Hampshire
Poultry Facility

Colovos Road
Durham, NH 03824

New Jersey
New Jersey Animal Health Diagnostic
  Lab
John Fitch Plaza CN 330
Trenton, NJ 08625

New Mexico
New Mexico Veterinary Diagnostic
  Lab
700 Camino De Salud NE
Albuquerque, NM 87106

New York
Cornell Duck Research Laboratories
Veterinary Diagnostic Lab
Box 217, Old Country Road
Eastport, L.I., NY 11941

New York State Veterinary College
Dept. of Avian and Aquatic Animal
  Medicine
Ithaca, NY 14853

Regional Veterinary Laboratory
88 Prince Street
Kingston, NY 12401

North Carolina
Western Animal Disease Diagnostic
  Laboratory
P.O. Box 279, Airport Road
Arden, NC 28704

North Carolina Dept. of Agriculture
Poultry Disease Diagnostic
  Laboratory - Veterinary Division
North Wilkesboro, NC 28659

Poultry Disease Diagnostic
   Laboratory
Box 476, Rockingham Street
Robbins, NC 27325

Poultry Disease Diagnostic
   Laboratory
Agriculture Bldg.
130 S. Port Road
Shelby, NC 28150

Ohio
Dept. of Veterinary Clinical Sciences
2578 Kenny Road
Columbus, OH 43210

Oklahoma
College of Veterinary Medicine
Oklahoma State University
Stillwater, OK 74074

Oregon
Oregon State Veterinary Diagnostic
   Laboratory
Oregon State University
Corvallis, OR 97331

Pennsylvania
Delaware Valley College of Science &
   Agriculture
Poultry Diagnostic Laboratory
Doylestown, PA 18901

Puerto Rico
Puerto Rico Animal Diagnostic
   Laboratory
P.O. Box E
Dorado, Puerto Rico 00646

South Carolina
Clemson University
Livestock Poultry Health Dept.
P.O. Box 218
Elgin, SC 29045

Tennessee
C. E. Kord Animal Disease
   Laboratory
P.O. Box 40627, Mel. Station
Nashville, TN 37204

Texas
Texas A & M University
Poultry Disease Laboratory
P.O. Box 84
College Station, TX 77843

Utah
Utah State University
Veterinary Diagnostic Laboratory
Logan, UT 84321

Utah Agricultural Experiment Station
Branch Veterinary Laboratory
P.O. Box 1068
Provo, UT 84603

Vermont
Animal Health Diagnostic Laboratory
University of Vermont
Animal Sciences — Bio-Research
   Complex
South Burlington, VT 05403

Virginia
Harrisonburg Regional Laboratory
Virginia Department of Agriculture
116 Reservoir Street
Harrisonburg, VA 22801

Virginia (continued)
Division of Animal Health
Regulatory Laboratory
Ivor, VA 23866

Division of Animal Health Laboratory
One N. 14th Street
Richmond, VA 23219

Poultry Diagnostic Laboratory
234 W. Shirley Avenue
Warrenton, VA 22186

Washington
H & N International Poultry
Health Research and Service
   Laboratory
3825 154th Avenue NE
Redmond, WA 98052-5300

West Virginia
State Federal Cooperative Animal
   Health Laboratory
State Capitol Bldg.
Charleston, WV 25305

Wisconsin
Regional Animal Health Laboratory
1418 LaSalle Avenue
Barron, WI 54812

Central Ave. Health Laboratory
6101 Mineral Point Road
Madison, WI 53705

Wyoming
Wyoming State Veterinary Laboratory
Box 950
Laramie, WY 82070

Cooperative Extension Service Offices by State

The Cooperative Extension Service is an excellent source of information on poultry and other agricultural enterprises. Extension personnel are located in county or area offices in each state as well as at agricultural colleges of the state universities. Poultry specialists are located at most state universities but not all of them. In some large poultry-producing states, specialists may be located in various areas of the state as well as on state university campuses. For information, contact the Cooperative Extension Service office nearest you.

Alabama
Auburn University
Auburn, AL 36849

Alaska
University of Alaska
Fairbanks, AK 99775

Arizona
University of Arizona
Tucson, AZ 85721

Arkansas
University of Arkansas
Fayetteville, AR 72701

California
University of California
Davis, CA 95616

University of California
Riverside, CA 92502

Colorado
Colorado State University
Fort Collins, CO 80523

Connecticut
University of Connecticut
Storrs, CT 06268

Delaware
University of Delaware
Newark, DE 19711

Florida
University of Florida
Gainesville, FL 32611

Georgia
University of Georgia
Athens, GA 30602

Hawaii
University of Hawaii
Honolulu, HI 96822

Idaho
University of Idaho
Moscow, ID 83843

Illinois
University of Illinois
Urbana, IL 61801

Indiana
School of Agriculture
Purdue University
West Lafayette, IN 47907

Iowa
Iowa State University of Science and
    Technology
201 Kildee Hall
Ames, IA 50011

Kansas
Kansas State University
Manhattan, KS 66506

Kentucky
Agricultural Science Center
University of Kentucky
Lexington, KY 40546

Louisiana
Louisiana State University
Baton Rouge, LA 70803

Maine
University of Maine
Orono, ME 04469

Maryland
University of Maryland
College Park, MD 20742

Massachusetts
University of Massachusetts
Amherst, MA 01003

Michigan
Michigan State University
East Lansing, MI 48824

Mississippi
Mississippi State University
Mississippi State, MS 39762

Missouri
University of Missouri
Columbia, MO 65211

Montana
Montana State University
Bozeman, MT 59717

Nebraska
University of Nebraska
Lincoln, NE 68583

Nevada
University of Nevada
College of Agriculture
Reno, NV 89557

New Hampshire
University of New Hampshire
Durham, NH 03824

New Jersey
Rutgers University
The University of New Jersey
New Brunswick, NJ 08903

New Mexico
New Mexico State University
Las Cruces, NM 88003

New York
Cornell University
Ithaca, NY 14853

North Carolina
North Carolina State University
Raleigh, NC 27695

North Dakota
North Dakota State University
Fargo, ND 58105

Ohio
Ohio State University
Columbus, OH 43210

Oklahoma
Oklahoma State University
Stillwater, OK 74078

Oregon
Oregon State University
Corvallis, OR 97331

Pennsylvania
Pennsylvania State University
University Park, PA 16802

Rhode Island
University of Rhode Island
Kingston, RI 02881

South Carolina
Clemson University
Clemson, SC 29631

South Dakota
South Dakota State University
Brookings, SD 57007

Tennessee
University of Tennessee
Institute of Agriculture
Knoxville, TN 37901

Texas
Texas A & M University
College Station, TX 77843

Utah
Utah State University
Logan, UT 84322

Vermont
University of Vermont
College of Agriculture
Burlington, VT 05405

Virginia
Virginia Polytechnic Institute and State
 University
Blacksburg, VA 24061

Washington
Washington State University
College of Agriculture
Pullman, WA 99164

West Virginia
West Virginia University
College of Agriculture
Morgantown, WV 26506

Wisconsin
University of Wisconsin
Madison, WI 53706

Wyoming
University of Wyoming
University Station
Laramie, WY 82071

Directory of Poultry Publications

A NUMBER of USDA and state Extension publications on poultry production and marketing are available and may be obtained by contacting the local county Extension Agent and state or area poultry Extension Specialists. The addresses of the Poultry specialists, by states, are given in a preceding section. A partial list of trade publications and books follows.

TRADE PUBLICATIONS

Broiler Industry, Watt Publishing Co., Mount Morris, IL 61054-1497

Egg Industry, Watt Publishing Co., Mount Morris, IL 61054-1497

Feedstuffs, 191 South Gary Ave., Carol Stream, IL 60188-2089

Poultry Digest, Watt Publishing Co., Mount Morris, IL 61054-1497

The Poultry Times, P.O. Box 1338, Gainesville, GA 30503

Turkey World, Watt Publishing Co., Mount Morris, IL 61054-1497

BOOKS

Commercial Chicken Production Manual. Mack O North, The AVI Publishing Co., Inc., Westport, CT 06880.

Poultry Production. Nesheim, Austic and Cord, Lea Febiger, Philadelphia, PA publisher.

Poultry Science and Production. Moreng and Avens, Reston Publishing Co., Inc., Reston, VA 22070.

Raising the Home Duck Flock. Dave Holderread, Garden Way Publishing, Pownal, VT 05261.

Raising Your Own Turkeys. Leonard S. Mercia, Garden Way Publishing, Pownal, VT 05261.

Sources of Supplies
and Equipment

MANY OF THE SUPPLIES and equipment needs for the small poultry flock can be found at the local feed store, hatchery, or other agricultural supply outlets. Some of the large mail-order houses such as Sears also handle agricultural supplies.

As an aid to locating sources of certain items not readily available in some areas, a partial list of supplies and equipment and possible sources is included below.

Space would not permit a complete listing of products or company names. There is no intent to recommend one source over another.

BAGS — DRESSED POULTRY

Dow Chemical Co. Flexible Packaging Sales, James Salvage Building, Midland, MI 48640
W.R. Grace and Co., Cryovac Division, Box 464, Duncan, SC 29334
Gordon Johnson Industries, 2519 Madison Avenue, Kansas City, MO 64108
Mobil Chemical Co., Plastics Div. Tech Center, Macedon, NY 14502

BANDS — IDENTIFICATION — LEG AND WING

Gey Band and Tag Co. Inc., P.O. Box 363, Norristown, PA 19401
National Band & Tag Co., 721 York Street, Newport, KY 41071

BROODERS

Beacon Industries, Inc., 100 Railroad Avenue, Westminster, MD 21157
Big Dutchman Cyclone Inc. , P.O. Box 1017, Holland, MI 49422-1017
Brower Equipment Co., P.O. Box 88, Houghton, IA 52631
Lyon Electric Co., 2765A Main Street, P.O. Box 3307, Chula Vista, CA 92011
Shenandoah Mfg. Co. Inc., P.O. Box 839, Harrisonburg, VA 22801

CAGES

Big Dutchman Cyclone, Inc., P.O. Box 1017, Holland, MI 49422-1017
Chore-Time Eqpt., Inc., P.O. Box 2000, Milford, IN 46542
Farmer Automatic, P.O. Box 39, Register, GA 30452

CANNIBALISM SUPPLIES AND EQUIPMENT

Gey Band and Tag Co., Inc., P.O. Box 363, Norristown, PA 19401
Kuhl Corporation, P.O. Box 26, Kuhl Road, Flemington, NJ 08822
Lyon Electric Co., 2765A Main Street, P.O. Box 3307, Chula Vista, CA 92011
NASCO, Fort Atkinson, WI 53538
National Band & Tag Co., 721 York Street, Newport, KY 41071

CAPONIZING INSTRUMENTS

NASCO, Fort Atkinson, WI 53538

CATCHING HOOKS

Beacon Industries, Inc., 100 Railroad Avenue, Westminster, MD 21157
Kuhl Corporation, P.O. Box 26, Kuhl Road, Flemington, NJ 08822
Shenandoah Mfg. Co., Inc., P.O. Box 839, Harrisonburg, VA 22801

DISINFECTANTS

Diversey Wyandotte Corp., 1532 Biddle Avenue, Wyandotte, MI 48192
Hess and Clark, Div. of Rhodia, Inc., 7th and Orange, Ashland, OH 44805
Salsbury Laboratories, Charles City, IA 50616
Vineland Laboratories, Inc., 2285 E. Landis Avenue, Vineland, NJ 08360

EGG BASKETS

Agri-Sales Associates, Inc., 212 Louise Avenue, Nashville, TN 37203
Beacon Industries, Inc., 100 Railroad Avenue, Westminster, MD 21157
Kuhl Corporation, P.O. Box 26, Kuhl Road, Flemington, NJ 08822

EGG CANDLERS — HAND

NASCO, Fort Atkinson, WI 53538
Brower Equipment Co., P.O. Box 88, Houghton, IA 52631

EGG CARTONS, CASES, AND FILLERS

Anderson Box Co., P.O. Box 1851, Indianapolis, IN 46206
Diamond Fiber Products, P.O. Box 627, Thorndike, MA 01079
Keyes Fibre, 3003 Summer Street, Stamford, CT 06905
Val-A Company, 700 W. Root Street, Chicago, IL 60609

EGG SCALES — SMALL

Beacon Industries, Inc., 100 Railroad Avenue, Westminster, MD 21157
Kent Company, 13145 Coronado Drive, Miami, FL 33181
Kuhl Corporation, P.O. Box 26, Kuhl Road, Flemington, NJ 08822
National Poultry Equipment Co., 1520B Standiford Avenue, Modesto, CA
    95350

EGG WASHERS

NASCO, Fort Atkinson, WI 53538

FEEDERS AND WATERERS

Beacon Industries, Inc., 100 Railroad Avenue, Westminster, MD 21157
Big Dutchman, P.O. 1017, Holland, MI 49422-1017
Shenandoah Mfg. Co., Inc., P.O. Box 389, Harrisonburg, VA 22801

INCUBATORS — SMALL

Brower Equipment, P.O. Box 88, Houghton, IA 52631
Lyon Electric Co., 2765A Main Street, P.O. Box 3307, Chula Vista, CA 92011
Petersyme Incubator Co., Gettysburg, OH 45328

KILLING CONES, KNIVES, AND OTHER DRESSING EQUIPMENT

Ashley Machine, Inc., 901 Carver Street, P.O. Box 2, Greensburg, IN 47240
Kuhl Corporation, P.O. Box 26, Kuhl Road, Flemington, NJ 08822
NASCO, Fort Atkinson, WI 53538
Nunnally Enterprises, Inc., 4550 North Star Way, Modesto, CA 95356

NESTS

Beacon Industries, Inc., 100 Railroad Avenue, Westminster, MD 21157
Big Dutchman Cyclone, Inc., P.O. Box 1017, Holland, MI 49422-1017
Shenandoah Mfg. Co., Inc., P.O. Box 839, Harrisonburg, VA 22801

POULTRY PICKERS AND SCALDERS

Ashley Machine, Inc., 901 Carver Street, P.O. Box 2, Greensburg, IN 47240
Kent Co., 13145 Coronado Drive, Miami, Florida 33181
Kuhl Corporation, P.O. Box 26, Kuhl Road, Flemington, NJ 08822
Pickwick Co., P.O. Box 756, 1120 Glass Road, Cedar Rapids, IA 52406

WATER WARMERS

Beacon Industries, Inc., 100 Railroad Avenue, Box 600, Westminster, MD 21157
Monoflo International, Inc., P.O. Box 2797, Winchester, VA 22601
NASCO, Fort Atkinson, WI 53538

WAX — DEFEATHERING

Continental Machine Co., P.O. Box 2936, 222 Myrtle Street, S.E., Gainesville, GA 30501
Pickwick Co., P.O. Box 756, 1120 Glass Road, Cedar Rapids, IA 52406

References

Aho, William A. and Talmadge, Daniel W., *Incubation and Embryology of the Chick*, Cooperative Extension Service, College of Agriculture and Natural Resources, University of Connecticut, Storrs, CT.

Ash, William J., *Raising Ducks*, Farmers Bulletin, No. 2215, USDA.

Caponizing Chickens, Leaflet 490, USDA.

Culling Hens, Farmers Bulletin No. 2216, Agricultural Research Service, USDA.

Duck Rations, Extension Stencil No. 25, Department of Poultry Science, Cornell University, Ithaca, NY.

Egg Grading Manual, Agricultural Handbook, No. 75, Consumer and Marketing Service, USDA.

Embryology and Biology of Chickens, The Vermont Extension Service, University of Vermont, Burlington, VT.

Farm Poultry Management, Farmers Bulletin No. 2197, Agricultural Research Service, USDA.

Brook, Munro, 4-H In-School Embryology Program, In-Vitro Demonstration, Univ. of Vermont Extension Service, Burlington, VT.

Hauver, W. E. and Kilpatrick, L., *Poultry Grading Manual*, Agricultural Handbook No. 31, USDA.

Hunter, J. M. and Scholes, John C., *Profitable Duck Management*, The Beacon Milling Co., Cayuga, NY.

Jordan, H. C. and Schwartz, L.D., *Home Processing of Poultry,* Penn State University, College of Agriculture Extension Service, University Park, PA.

Marsden, Stanley J., *Turkey Production,* Agricultural Handbook No. 393, Agricultural Research Service, USDA.

Mercia, L. S., *Drawing Poultry,* 4-H Publication, Vermont Extension Service, University of Vermont, Burlington, VT.

Mercia, L. S., *Killing, Picking and Cooling Poultry,* 4-H Publication, Vermont Extension Service, University of Vermont, Burlington, VT.

Mercia, L. S., *Lighting Replacements and Layers,* Vermont Extension Service, University of Vermont, Burlington, VT.

Mercia, L. S., *The Family Laying Flock,* Brieflet 1218, Vermont Extension Service, University of Vermont, Burlington, VT.

Moyer, D. D., *Virginia Turkey Management,* Publication 302, Extension Division, Virginia Polytechnic Institute, Blacksburg, VA.

Orr, H. L., *Duck and Goose Raising,* Publication 532, Ontario Ministry of Agriculture and Food, Department of Animal and Poultry Science, Ontario Agriculture College, Guelph, Ontario, Canada.

Ota, Hajime, *Housing and Equipment for Laying Hens,* Misc. Publication 728, Agricultural Research Service, USDA.

Poultry Management and Business Analysis Manual for the 80's, Bulletin 566 (Revised), Cooperative Extension Services, Extension Poultrymen in New England.

Raising Geese, Farmers Bulletin, No. 2251, Northeastern Region Agricultural Research Service, USDA.

Raising Livestock on Small Farms, Farmers Bulletin No. 2224, USDA.

From Egg to Chick, A 4-H Manual of Embryology and Incubation, Northeast Cooperative Publications.

Ridlen, S. F., and Johnson, H. S., *4-H Poultry Management Manual — Unit 1*, Cooperative Extension Service, University of Illinois, Urbana, IL.

Talmadge, Dan W., *Brooding and Rearing Baby Chicks,* Published by the Cooperative Extension Services of the Northeast.

Warren, Richard, *4-H Laying Flock Management,* 4-H Circular 79, Published by the Cooperative Extension Services in New England.

TABLE A
GENERAL FEED REQUIREMENTS
FOR DIFFERENT TYPES OF BIRDS
AT VARIOUS AGES

Type of Bird		Type Feed	Percent Protein	Lbs./ 100 Birds White Egg	Brown Egg
Chickens:					
Pullets	0–8 wks.	Starter	20	280	364
	8–12 wks.	Grower	17	308	392
	12–22 wks.	Grower or Developer	14	1064	1289
Layers	22 + wks.	Layer	17	23	25
Broilers	0–4 wks.	Starter	22	200	
or	5–7 wks.	Grower	20	500	
Fryers	7–8 wks.	Finisher	18	100	
Roasters	0–4 wks.	Starter	22	200	
	4–13 wks.	Grower	16	1200	
	13–market	Finisher	14	1000	
Turkeys:					
Small	0–7 wks.	Starter	28	600	
	7–18 wks.	Grower	20	3000	
Large	0–8 wks.	Starter	28	1000	
	8–16 wks.	Grower	20	3000	
	16–24 wks.	Finisher	14	3500	
Ducks:					
(White Pekin)	0–2 wks.	Starter	22-24	214	
For market	2–4 wks.	Grower	18-20	510	
	4–8 wks.	Finisher	16-17	1410	
For Breeders					
	0–2 wks.	Starter	22-24	214	
	2 wks.– maturity	Grower	18-20	*	
Geese:					
(White Emden	0–3 wks.	Starter	20-22	525	
or Chinese)	3–14 wks.	Grower	15	5600	
	14 wks. – market (24–30 wks.)	Grower	15		

*Ducks and particularly geese are good foragers. From approximately 3 weeks of age with good forage available, savings of up to 35% in feed consumption can be realized.

TABLE B
A "NATURAL" LAYING HEN DIET
To be fed as all-mash to medium-size layers kept in floor pens

	AMOUNT (LBS.)	
INGREDIENT	100 LBS.	PERTON
Yellow corn meal	60.00	1200
Wheat middlings	15.00	300
Soybean meal (dehulled)	8.00	160
Maine herring meal (65%)	3.75	75
Meat & bone meal (47%)	1.00	20
Skim milk, dried	3.00	60
Alfalfa leaf meal (20%)	2.50	50
Iodized salt	0.40	8
Limestone, grd. (38% Ca)	6.35	127
TOTALS	100.00	2000

CALCULATED ANALYSIS	NATURAL LAYING HEN DIET (per pound)	RECOMMENDED (NECC)* NUTRITIONISTS (per pound)
Metabolizable energy Cal./lb.	1252	1292
Protein	16.07%	16.00%
Lysine	0.79%	0.74%
Methionine	0.31%	0.29%
Methionine & cystine	0.55%	0.54%
Fat	3.67%	3.33%
Fiber	3.15%	2.51%
Calcium	2.77%	2.75%
Total phosphorus	0.53%	0.50%
Available phosphorus	0.44%	0.42%

VITAMINS (UNITS OR MGS./LB.)

Vitamin A activity (U.S.P. units/lb.)	5112.00	5290.00
Vitamin D (I.C.U.)		1000.00
Riboflavin (mg.)	1.36	1.38
Pantothenic acid (mg.)	3.89	4.05
Choline Chloride (mg.)	411. 00	500.00
Niacin (mg.)	17.46	16.95

The birds *must* receive *direct* sunlight to enable them to synthesize vitamin D. Unforti-fied cod liver oil can be fed in place of sunlight to supply vitamin D. The amount of cod liver oil would depend upon the potency of the oil — the need is for 1000 (I.C.U.)** per pound of feed. (Prepared by Dr. Richard Gerry, Department of Animal & Veterinary Sciences, University of Maine, Orono, Maine)

*New England College Conference nutritionists **International Chick Unit

TABLE C
POULTRY MANURE INFORMATION

MANURE PRODUCTION

Layers: 25 pounds per 100 per day with normal drying.

Four-tenths of a cubic foot per 100 per day.

Moisture content 75-80% as defecated.

Pollution load—7 to 12 layers equal one human.

Weight of a cubic foot of poultry manure at 70% moisture approximately 65 pounds.

Broilers: Manure and litter per 100 birds

8 weeks equals 400 pounds or 11.2 cubic feet at 25% moisture.

FERTILIZER VALUE OF POULTRY MANURE*

			POUNDS PER TON	
	MOISTURE %	NITROGEN	PHOSPHORUS	POTASH
Fresh Manure	75	29	10	8
Stored Manure	63.9	24	13	16
Broiler Litter	18.9	72	25	30
Layer Litter	22.1	50	23	36
Slurry (liquid)	92	22	12	7

When nitrogen is worth 25¢ per pound, phosphorus 15¢ per pound, and potash 8¢ per pound, the values per ton of poultry manure are:

Fresh Manure	$9.39
Stored Manure	9.23
Broiler Litter	24.15
Layer Litter	14.86

Other elements of plant food contained in poultry manure include calcium, magnesium, copper, manganese, zinc, chlorine, sulphur, and boron.

*Based on information from Univ. of New Hampshire *Bulletin 444 — Farm Manure* , 1966.

Glossary of Terms

Abdomen. Area between the keel and pubic bones.

Air cell. Air space usually found in the large end of the egg.

Albumen. The white of the egg.

American breeds. Those breeds developed in America and having common characteristics such as yellow skin, nonfeathered shanks, red earlobes. All lay brown eggs except the Lamonas, which produce white eggs.

Axial feather. The short wing feather between the primaries and secondaries.

Baby chick. Newly hatched chick before it has been fed or watered.

Bantam. Diminutive fowl. Some are distinct breeds, others are miniatures of large breeds.

Beak. Upper and lower mandibles of chickens, turkeys, pheasants, peafowl, etc.

Bean. Hard protuberance on the upper mandible of waterfowl.

Bill. The upper and lower mandibles of waterfowl.

Blood spot. Blood in an egg caused by a rupture of small blood vessels, usually at the time of ovulation.

Breast. The forward part of the body between the neck and the keel bone.

Breast blister. Enlarged, discolored area or sore in the area of the keel bone.

Breed. A group of fowl related by ancestry and breeding true to certain characteristics such as body shape and size.

Broiler-fryer. Young chickens under twelve weeks of age of either sex that are tender-meated with soft, pliable, smooth-textured skin and flexible breastbone cartilage.

Brooder. Heat source for starting young birds.

Broody. Maternal instinct causing the female to set or want to hatch eggs.

Candle. To determine the interior quality of an egg through the use of a special light in a dark room.

Cannibalism. A vice usually manifested by toe picking, vent picking, or feather pulling.

Capon. A castrated male fowl having undeveloped comb and wattles, and longer hackles, saddle, and tail feathers than the normal male.

Chalazae. White, twisted, ropelike structures that tend to anchor the egg yolk in the center of the egg, by their attachment to the layers of thick albumen.

Chalaziferous layer. Thin layer of thick white surrounding the yolk, continuous with the chalazae.

Cloaca. The common chamber or receptacle for the digestive, urinary, and reproductive systems.

Cock. A male bird over twelve months of age.

Cockerel. A male bird under twelve months of age.

Coccidiostat. A drug used to control or treat coccidiosis disease.

Comb. The fleshy prominence on the top of the head of fowl.

Crop. An enlargement of the gullet where food is stored and prepared for digestion.

Crossbred. The first generation resulting from crossing two different breeds or varieties, or the crossing of first generation stock with first generation stocks resulting from the crossing of other breeds and varieties.

Cull. A bird not suitable to be in a laying or breeding pen or not suitable as a market bird.

Culling. The act of removing unsuitable birds from the flock.

Debeak. To remove a part of the beak to prevent feather pulling or cannibalism.

Drake. Male duck.

Dub. To trim the comb and wattles close to the head.

Duck. Any member of the family *Anatidae* and specifically a female.

Duckling. The young of the family *Anatidae*.

Ear lobe. Fleshy patch of skin below ear. It may be red, white, blue, or purple, depending upon the breed of chicken.

Embryo. A young organism in the early stages of development, as before hatching from the egg.

Face. Skin around and below the eyes.

Flight feathers. Primary feathers of the wing, sometimes used to denote the primaries and secondaries.

Follicle. Thin, highly vascular ovarian tissue containing the growing ovum.

Footcandle. The density of light striking each and every point on a segment of the inside surface of an imaginary one-foot radius sphere with a one-candle-power source at the center.

Fowl. Term applied collectively to chickens, ducks, geese, etc., or the market class designation for old laying birds.

Gander. Male goose.

Germinal disc, or blastodisc. Site of fertilization on the egg yolk.

Gizzard. Muscular stomach. Its main function is grinding food and partial digestion of proteins.

Goose. The female goose as distinquished from the gander.

Gosling. A young goose of either sex.

Grit. The hard, insoluble materials fed to birds to provide a grinding material in the gizzard.

Gullet, or esophagus. The tubular structure leading from the mouth to the glandular stomach.

Hackle. Plumage on the side and rear of the neck of fowl.

Hen. A female fowl more than twelve months of age.

Hock. The joint of the leg between the lower thigh and the shank.

Horn. Term used to describe various color shadings in the beak of some breeds of fowl such as the Rhode Island Red.

Hover. Canopy used on brooder stoves to hold heat near the floor when brooding young stock.

Infundibulum, or funnel. Part of the oviduct that receives the ova (egg yolks) when they are ovulated. Fertilization of the egg takes place in the infundibulum.

Isthmus. Part of the oviduct where the shell membranes are added during egg formation.

Keel bone. Breast bone or sternum.

Litter. Soft, absorbent material used to cover floors of poultry houses.

Magnum. Part of the oviduct that secretes the thick albumen or white during the process of egg formation.

Mandible. The upper or lower bony portion of the beak.

Mash. A mixture of finely ground grains.

Meat spots. Generally, blood spots that have changed in color due to chemical action.

Molt. To shed old feathers and regrow new ones.

Oil sac, or uropygial gland. Large oil gland on the back at the base of the tail — used to preen or condition the feathers.

Ova. Round bodies (yolks) attached to the ovary. These drop into the oviduct and become yolk of the egg.

Oviduct. Long, glandular tube where egg formation takes place; leads from the ovary to the cloaca. It is made up of the funnel, magnum, isthmus, uterus, and vagina.

Pendulous crop. Crop that is usually impacted and enlarged and hangs down in an abnormal manner.

Plumage. The feathers making up the outer covering of fowls.

Poult. A young turkey.

Primaries. The long, stiff flight feathers at the outer tip of the wing.

Pubic bones. The thin terminal portion of the hip bones that form part of the pelvis. Used as an aid in judging productivity of laying birds.

Pullet. Female chicken less than one year of age.

Recycle, or force molt. To force into a molt with a cessation of egg production.

Relative humidity. The percentage of moisture saturation of the air.

Replacements. Term used to describe young birds that will replace an old flock.

Roasters. Young chickens of either sex, usually three to five months of age, that are tender-meated with soft, pliable, smooth-textured skin and with a breastbone cartilage, somewhat less flexible than the broiler-fryer.

Roost. A perch on which fowl rest or sleep.

Secondaries. The large wing feathers adjacent to the body, visible when the wing is folded or extended.

Sex-linked. Any inherited factor linked to the sex chromosomes of either parent. Plumage color differences between the male and female progeny of some crosses is an example of sex-linkage. Useful in sexing day-old chicks.

Shell membranes. The two membranes attached to the inner eggshell. They normally separate at the large end of the egg to form air cells.

Slip. A male from which all of both testicles were not removed during the caponizing operation.

Snood. Fleshy appendage on the head of a turkey.

Sperm, or spermatozoa. The male reproductive cells capable of fertilizing the ova.

Spur. The stiff, horny process on the legs of some birds. Found on the inner side of the shanks.

Standard-bred. Conforming to the description of a given breed and variety as described in the American Standard of Perfection.

Sternum. The breast bone or keel.

Stigma. The suture line or nonvascular area where the follicle ruptures when the mature ovum is dropped.

Strain. Fowl of any breed usually with a given breeder's name and which has been reproduced by closed-flock breeding for five generations or more.

Testicles, or testes. The male sex glands.

Trachea, or windpipe. That part of the respiratory system that conveys air from the larynx to the bronchi and to the lungs.

Undercolor. Color of the downy part of the plumage.

Uterus. The portion of the oviduct where the thin white, the shell, and shell pigment are added during egg formation.

Vagina. Section of the oviduct that holds the formed egg until it is laid.

Vent, or anus. Is the external opening from the cloaca.

Vitelline membrane. Thin membrane that encloses the ovum.

Wattles. The thin pendant appendages at either side of the base of the beak and upper throat, usually much larger in males than in females.

Windpuffs. Air trapped under the outer skin as a result of rupturing the air sacs during caponization.

Xanthophyll. One of the yellow pigments found in green plants, yellow corn, fatty tissues, and egg yolks.

Yolk. Ovum, the yellow portion of the egg.

Index

roosts for, 25-27; site selection of, 8-11; space requirements for, 12-13; types of, 7-8; ventilation for, 15-17; waterers for, 22-23

Poultry meat: processing of, 159-61

Poultry Specialist, 50, 112

Poults: brooding, 118-19; debeaking of, 126; definition of, 227; feeding of, 118; selection of, 112-13; separation from older turkeys, 128. *See also* Turkeys

Predators: poultry losses from, 48; problem of, 195-96

Protein: in diet, 58

Protein-supplement-plus-grain system: of feeding, 120

Pubic bones: of laying birds, 67-69

Pullets: cost of growing pullets, 55; definition of, 33; diet for, 45-46; egg laying time, 34-37; facilities for, 48-49; feeding of, 44-45; lighting program for, 52-54; purchasing of, 34-35. *See also* Chicks

Pullorum. *See* Salmonella pullorum

R

Raccoons: predator problem of, 196

Range: for geese, 150-51

Range feeder: use of, 124-25

Range rearing: confinement rearing versus, 132; feeding in, 48; of turkeys, 121-25; versus confinement rearing, 47-49

Range shelter: for turkeys, 125; use of, 48-49

Rearing: of waterfowl, 139-41

Records: keeping of, 76-77

Recycle: definition of, 227

Recycle. *See* Force molt

Red mite: parasite of, 191-92

Replacements: definition of, 227

Reproduction: reproductive system of hens, 92-93

Rhode Island Red: broodiness of, 87; mating of, 79; purpose of, 34-35; sex linkage of, 90

Rickets: disease of, 194

Roaster. *See also* Meat birds

Roasters: cost of production, 109; eviscerating of, 165-70; facilities for, 48-49; growth rate/feed consumption for, 122; raising of, 2-3, 100; water consumption for, 123

Rodents: control of, 66, 128; problem of, 195-96

Roosting: problems of, 101

Roosts: use of, 25-27, 41, 119-20

Rouen duck, 135; characteristics of, 136

Roundworm (large): parasite of, 189

S

Salmonella pullorum: disease of, 184-185; test for, 79, 134

Salmonella typhoid: test for, 79, 134

Scalding tank: use of, 160

Scaly leg mite: parasite of, 192-93

Schackles: use of, 159-60

Scratch feed: for laying birds, 59-60

Sebastipole goose, 137; characteristics of, 138-39

Semen: used for artificial insemination, 131

Semiscald: method of, 163-64

Sex: identification of chicks, 89-90

Sex-Sal layer: purpose of, 35

Sexing: of ducklings/goslings, 150

Sexual maturity: delaying of, 46-47

Shanks: bleaching of, 70-71; of laying birds, 68

Shell: of an egg, 153-54; color of, 33-34; quality of, 81. *See also* Egg

Shell membranes: of an egg, 153

Shell window method: embryo development and, 93-94

Shelters: in range rearing, 48-49

Singeing: method of, 164

Single Comb White Leghorn: combs of, 75

Sinusitus (infectious): test for, 134

Skunks: problem of, 196

Slips: characteristics of, 107; gonads and, 169

Space: for meat birds, 100; for poultry housing, 12-13

Specs: for anti-picking control, 51

Sperm: fertilization of turkey eggs, 130-31

State Department of Agriculture, 79

Step-down program: use of, 54

Still-air incubator: use of, 82, 89

Storage: of hatching eggs, 82

Strain: definition of, 228

Subscald: method of, 162-63

Sunporch. *See* Porch

T

Tapeworm: parasite of, 190-91

Temperature: for brooding, 43-44; water consumption of layers and, 64

Thermometer: for scalding, 160-61

Toe clipping: of turkeys, 128

Toe picking: problem of, 50

Toms: lighting needs for, 129; turkey mating, 129-30

Toulouse goose: characteristics of, 137

Tropical chicken flea: parasite of, 192

Trough-type feeders: use of, 124

Troughs: use of, 141

Trussing: method of, 170

Turkey: breeds of, 112; cost of raising heavy roaster turkeys, 111; feeding of, 118; housing requirements for, 113-16; incubation period for, 87; raising of, 3. *See also* Poults

Turkey breeders: feed consumption of, 130; feeding of, 132-33

Turkeys: broodiness of, 131-32; care of, 128; diet for, 45; egg production of, 130; eviscerating of, 165-70; feeding of, 120-21; grooming of, 125-28; management recommendations for, 128; range rearing

Turkeys: broodiness of *(continued)*
of, 121-25; range shelter for, 125; selection of breeders, 129; separation from poults, 128. *See also* Meat birds

U

United States: egg quality standards, 156; weight classes for consumer grades or shell eggs, 157
United States Department of Agriculture, 155, 157

V

Vaccination: for poultry diseases, 49-50
Vent: color changes of, 69-70; of laying birds, 67-69
Ventilation: of incubator, 83; for poultry house, 66; for poultry housing, 15-17
Virus hepatitis: disease of, 188-89
Vitamin A deficiency: problem of, 195
Viteline membrane: of an egg, 154

W

Water: broiler consumption of, 99; consumption of, 63; consumption of roaster turkeys,123; for meat birds, 100; for poults, 118
Water cooling: method of, 171-72
Waterers: for brooding chicks, 40; for laying birds, 63; use of, 22-23, 141-42
Waterfowl: brooding of, 139-41; diseases of, 186-89; eviscerating of, 165-70; feathers

of, 151-52; feeding of, 141-42; housing for, 144-46; raising of, 4, 135, 139. *See also* Meat birds
Waterfowl. *See* specific breeds by name
Waterfowl: sex differentiation of, 148-49
Wattles: after caponization, 104; of laying birds, 67-69
Weasels: predator problem of, 196
Weeds: geese as weeders, 151
Weight (weighted) blood cup: use of, 161
White Calls duck, 135
White Chinese Emden goslings: growth rate/ feed consumption of, 144
White Cornish: caponizing of, 104
White Cornish. *See* White Plymouth Rock
White diarrhea. *See* Salmonella Pullorum
White Leghorns: dubbing of, 76
White Pekin duck, 135; characteristics of, 135-36
White Plymouth Rock: breeding of, 97; caponizing of, 104
White Rock Crosses: caponizing of, 104
Wing clipping: of turkeys, 127
Wings: stages of molt, 72
Wood Duck, 135
Wyandottes: broodiness of, 87

X
Xanthophyl. *See* Pigmentation

Y
Yards: use of, 48-49
Yolk: of an egg, 153-54